Low Pressure Boilers
Second Edition

Workbook

AMERICAN TECHNICAL PUBLISHERS, INC.
HOMEWOOD, ILLINOIS 60430-4600

Frederick M. Steingress

© 2005 by American Technical Publishers, Inc.
All rights reserved

1 2 3 4 5 6 7 8 9 – 05 – 9 8 7 6 5 4 3 2 1

Printed in the United States of America

ISBN 0-8269-4351-9

Contents

1 Boiler Operation Principles 1

2 Boiler Fittings 7

3 Feedwater System 17

4 Steam System 25

5 Fuel System 33

6 Draft System 49

7 Boiler Water Treatment 55

8 Boiler Operation Procedures 61

9 Hot Water Heating Systems 69

10 Cooling Systems 77

11 Boiler Operation Safety 85

12 Test 1 89
Test 2 97
Test 3 105
Test 4 111
Test 5 117
Test 6 123
Test 7 131

Introduction

Low Pressure Boilers Workbook is designed to reinforce information presented in *Low Pressure Boilers*, 2nd Edition. The textbook is used as a reference to complete the learning activities in the workbook. Each chapter in the workbook covers information in the corresponding chapter in the textbook. Reference text pages of the textbook are cited at the beginning of each workbook chapter. Chapter 12 – Testing contains seven separate tests that can be used as a comprehensive review. The questions in Chapter 12 are similar to the type of questions found on a typical boiler operator licensing examination.

Answering Chapter Questions

The question types used in the workbook include true-false, multiple choice, matching, and essay. For true-false questions, circle the letter T if the statement is true, or circle the letter F if the statement is false. For multiple choice questions, place the letter of the correct answer in the answer blank next to the question. For matching questions, place the letter of the correct corresponding answer in the answer blank next to the question. For essay questions, write the answers on a separate sheet of paper and submit to the instructor.

ASME Code Symbol Stamps

At the bottom of the first page in Chapters 1 through 11, a new ASME Code symbol stamp and its meaning is introduced. This allows progressive learning of the symbol stamps used on various boiler room equipment. Chapter 12 checks for understanding of the symbol stamps presented throughout the workbook.

Related Information

Information presented in *Low Pressure Boilers*, 2nd Edition and the *Low Pressure Boilers Workbook* addresses common boiler operation topics. Additional information related to boiler operation/stationary engineering is available in other American Tech products. To obtain information about these products, visit the American Tech web site at www.go2atp.com

The Publisher

Boiler Operation Principles
Text Reference Pages 1 – 30

Name_____ Date_____

True-False

(T) F   1. Steam is formed when water is heated to its boiling point.

T (F)   2. A boiler is an open metal container in which water is heated to produce steam or heated water. *CLOSED*

(T) F   3. Water begins to boil at approximately 212°F.

(T) F   4. In a scotch marine boiler, the gases of combustion pass through tubes that are surrounded by water.

(T) F   5. Oxygen is needed to burn fuel.

(T) F   6. Condensate is returned to the boiler for reuse.

T (F)   7. The steam system supplies water to the boiler. *FEED WATER SYSTEM*

(T) F   8. Steam leaves the boiler through the main steam line.

T (F)   9. Steam that has given up its heat and turned back to water is boiler water. *CONDENSATE*

(T) F   10. Heat is generated in a boiler by the combustion of a fuel such as gas, fuel oil, or coal.

(T) F   11. An internal furnace is a furnace in a boiler that is surrounded by heating surface.

(T) F   12. Feedwater is water that is treated for use in the boiler.

T (F)   13. Heat always flows from a material having a lower temperature to a material having a higher temperature. *HIGHER TO LOWER*

(T) F   14. Latent heat is heat identified by a change of state and no temperature change of the substance.

(T) F   15. Conduction is heat transfer that occurs when molecules in a material are heated and the heat is passed from molecule to molecule through the material.

T (F)   16. Approximately 212 lb of air is required to burn a pound of fuel. ?

(T) F   17. The ASME Code governs boiler design, material, method of construction, inspection, and quality assurance.

ASME CODE SYMBOL STAMP   HEATING BOILER

T **F** 18. A low pressure boiler is a boiler that has a MAWP over 15 psi.

T F 19. A package boiler is a self-contained unit that is preassembled at the factory.

T **F** 20. A firebox boiler is a cylindrical scotch marine boiler.

Multiple Choice

__D__ 1. _____ is necessary to generate steam in a boiler.
- A. A container
- B. Water
- C. Heat
- D. all of the above

__B__ 2. The _____ is the part of the boiler with water on one side and heat on the other.
- A. furnace volume
- B. heating surface
- C. fire side
- D. water side

__A__ 3. A _____ boiler has heat and gases of combustion that pass through tubes surrounded by water.
- A. firetube
- B. watertube
- C. cast iron
- D. straight-tube

__B__ 4. A _____ boiler has water in the tubes and heat and gases of combustion passing around the tubes.
- A. firetube
- B. watertube
- C. cast iron
- D. firebox

__C__ 5. _____ are used in boilers to direct the gases of combustion over the boiler heating surface.
- A. Combustion controls
- B. Firetubes
- C. Baffles
- D. Zone controls

__D__ 6. Sensible heat is heat that _____.
- A. involves a change of state
- B. has no temperature change of the substance
- C. changes ice to water
- D. can be measured with a thermometer

__B__ 7. A _____ is anything that transfers heat from one substance to another without allowing the materials to mix.
   A. thermometer
   B. heat exchanger
   C. thermostat
   D. regulator

__C__ 8. In the _____, air mixes with fuel and burns.
   A. tube sheet
   B. watertubes
   C. combustion chamber
   D. breeching

__A__ 9. _____ combustion is combustion that occurs when fuel is burned using only the theoretical amount of air.
   A. Perfect
   B. Incomplete
   C. Complete
   D. Partial efficient

__B__ 10. _____ air is air that controls the combustion rate, which determines the amount of fuel burned.
   A. Excess
   B. Primary
   C. Secondary
   D. Pilot

__C__ 11. The breaking up of fuel into small particles to maximize contact of fuel with air for combustion is _____.
   A. steam priming
   B. convection loading
   C. atomization
   D. discharge shaping

__B__ 12. The _____ system provides the air necessary for combustion.
   A. feedwater
   B. draft
   C. steam
   D. fuel

__C__ 13. The _____ is the highest pressure in pounds per square inch at which a boiler can safely be operated.
   A. NOWL
   B. ASME
   C. MAWP
   D. LWFC

4 LOW PRESSURE BOILERS WORKBOOK

__A__ 14. A _____ boiler does not use tubes.
- A. cast iron
- B. scotch marine
- C. watertube
- D. firebox

__C__ 15. To increase efficiency of the boiler, _____.
- A. more fuel is added
- B. firetubes are decreased in size
- C. the heating surface is increased
- D. all of the above

__D__ 16. The four systems necessary to operate a boiler are _____, _____, _____, and _____.
- A. combustion; draft; steam; boiler water
- B. water; steam; combustion; stoker
- C. boiler water; fuel; draft; condensate
- D. feedwater; fuel; draft; steam

__A__ 17. _____ pressure is the pressure caused by the weight of air surrounding the earth.
- A. Atmospheric
- B. Thermal
- C. Conductive
- D. Absolute

__D__ 18. During operation, a boiler _____.
- A. holds water
- B. collects the steam that is produced
- C. transfers heat to the water to produce steam
- D. all of the above

__D__ 19. Water turns to steam at _____ °F.
- A. 100
- B. 150
- C. 200
- D. 212

__D__ 20. _____ boilers have sections that can be assembled on-site to produce the boiler capacity required.
- A. Firetube
- B. Industrial watertube
- C. Scotch marine
- D. Cast iron

__D__ 21. The amount of energy required to raise the temperature of 1 lb of water 1°F is one _____.
   A. NOWL
   B. therm
   C. psia
   D. Btu

__D__ 22. Heat is transferred by _____.
   A. conduction
   B. convection
   C. radiation
   D. all of the above

__A__ 23. A(n) _____ is a component directly attached to the boiler that is required for the operation of the boiler.
   A. fitting
   B. accessory
   C. fuel oil tank
   D. condensate return tank

__A__ 24. The nonflammable material used to insulate the outer surface of the boiler from heat is _____.
   A. refractory
   B. steam bridge
   C. aquastat
   D. tube bank

__D__ 25. A firetube boiler that has a long cylindrical shape is a _____ boiler.
   A. firebox
   B. cast iron
   C. membrane watertube
   D. scotch marine

Boiler Systems

__A__ 1. Steam system
__C__ 2. Boiler
__B__ 3. Feedwater system
__E__ 4. Draft system
__D__ 5. Fuel system

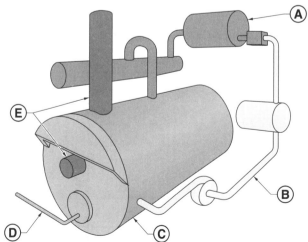

Scotch Marine Boiler

<u>B</u>   1. Tubes
<u>A</u>   2. Tube sheet
<u>D</u>   3. Gases of combustion
<u>C</u>   4. Internal furnace

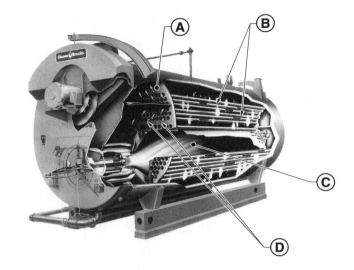

Steam Heating System

<u>D</u>   1. Condensate receiver tank
<u>G</u>   2. Main steam line
<u>E</u>   3. Feedwater pump
<u>I</u>   4. Steam header
<u>C</u>   5. Branch lines
<u>H</u>   6. Main steam stop valve
<u>A</u>   7. Heating unit
<u>B</u>   8. Steam trap
<u>F</u>   9. Boiler

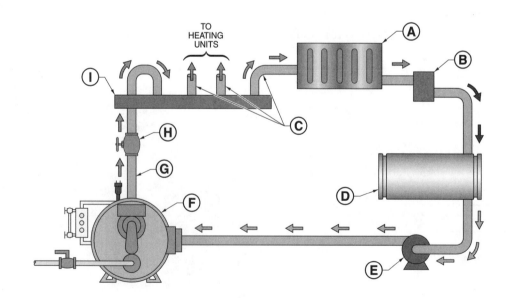

Boiler Fittings

Text Reference Pages 31 – 60

2

Name _____   Date _____

True-False

- (T) F   1. Boiler fittings are components attached directly to the boiler.
- T (F)   2. Safety valves on steam boilers are designed to open slowly.
- T (F)   3. The safety valve should *not* chatter to prevent damage to the disc.
- T (F)   4. Safety valves are tested only by performing a *hand* try lever test.
- T (F)   5. A safety valve try lever test is performed with the boiler at 5% of the safety valve set pressure.   *5 lbs of pressure*
- (T) F   6. The steam pressure gauge must be connected to the highest part of the steam side of the boiler.
- T (F)   7. A siphon is connected between the boiler and the steam pressure gauge to prevent water from entering the Bourdon tube. *steam*
- (T) F   8. A slow gauge reads less pressure than is actually in the boiler.
- (T) F   9. Live steam allowed to enter the Bourdon tube of a steam pressure gauge will damage the gauge.
- (T) F   10. The water column should be blown down once a shift.
- T (F)   11. The water column is used to indicate the water level in the blowdown tank.   *boiler*
- (T) F   12. With a NOWL in a boiler, the gauge glass is approximately half full.
- (T) F   13. The purpose of the water column is to slow down the movement of the boiler water to obtain a more accurate reading of the water level in the gauge glass.
- (T) F   14. In low pressure steam boilers, two methods of determining water level are try cocks and gauge glasses.
- T (F)   15. Gauge glasses are calibrated in pounds per square inch.   *steam pressure gauge*
- T (F)   16. With a NOWL, steam is discharged from the bottom try cock when it is opened.
- (T) F   17. With a NOWL, the middle try cock discharges a mixture of water and steam when opened.

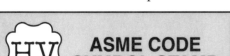

ASME CODE SYMBOL STAMP   SAFETY RELIEF VALVE

T	(F)	18. With a NOWL, water is discharged from the top try cock when opened.
(T)	F	19. If the top line to the gauge glass is closed or clogged, the glass will fill with water.
T	(F)	20. If the bottom line to the gauge glass is closed or clogged, the glass will be empty. *WATER STATIONARY*
T	(F)	21. To ensure the lines to the gauge glass are clear, it is recommended that the gauge be blown down at least once a month. *DAILY once a shift*
T	(F)	22. The bottom blowdown line must be connected to the highest part of the steam side of the boiler.
(T)	F	23. A high surface tension on boiler water can lead to foaming.
T	(F)	24. Surface impurities are removed from the boiler by using the ~~bottom~~ *surface* blowdown valves.
(T)	F	25. Boilers equipped with a quick-opening valve and a slow-opening valve must have the quick-opening valve closest to the boiler.
T	(F)	26. Only watertube boilers have bottom blowdown valves.
(T)	F	27. A high water level in the boiler can lead to carryover.
T	(F)	28. When dumping a boiler, there should be ~~maximum~~ *NO* steam pressure in the boiler to force water from the boiler.
T	(F)	29. To correct a high water level condition in the boiler, the operator must open the ~~surface~~ *BOTTOM* blowdown valve.
(T)	F	30. Bottom blowdown valves are used when dumping the boiler.
(T)	F	31. A fusible plug is the last warning the operator has of a dangerous low water condition in the boiler.
(T)	F	32. The core of the fusible plug melts at approximately 450°F.
(T)	F	33. The boiler vent is located at the highest part of the steam side of the boiler.
T	(F)	34. The safety valve should be used as a boiler vent for faster blowdown.
(T)	F	35. The boiler vent should be open when filling the boiler with water.
(T)	F	36. The pressure control is located at the highest part of the steam side of the boiler.
(T)	F	37. The pressure control must be mounted in a vertical position to ensure accurate operation.
T	(F)	38. A burner should always start up in ~~high~~ *low* fire to ensure enough fuel for ignition.
T	(F)	39. Burners using fuel oil or gas should be adjusted so they are OFF for longer periods than they are ON to save fuel and keep furnace temperatures lower.
(T)	F	40. The operating range is obtained by adjusting the cut-in pressure setting and differential pressure setting on the pressure control.

Multiple Choice

___B___ 1. The _____ is the most important valve on a boiler.
   A. main steam stop valve
   B. safety valve
   C. automatic nonreturn valve
   D. feedwater stop valve

___B___ 2. The MAWP on a low pressure steam boiler is _____ psi.
   A. 10
   B. 15
   C. 20
   D. 30

___C___ 3. Total force is equal to _____.
   A. area times diameter
   B. area times distance
   C. area times pressure
   D. MAWP times pressure

___D___ 4. The ASME code states that boilers with over _____ square feet of heating surface must have two or more safety valves.
   A. 200
   B. 300
   C. 400
   D. 500

___D___ 5. The area of a safety valve 4″ in diameter is _____ square inches.
   A. 2.3562
   B. 3.1416
   C. 6.2832
   D. 12.5664

DIAMETER² × .7854

___D___ 6. The only valve permitted between the safety valve and the boiler is the _____ valve.
   A. os&y gate
   B. os&y globe
   C. automatic nonreturn
   D. no valves are permitted between the boiler and safety valve

___B___ 7. The range of the pressure gauge should be _____ times the MAWP of the boiler.
   A. 1 to 2
   B. 1½ to 2
   C. 2 to 3
   D. 2½ to 3

8. In most states, _____ safety valves are permitted on steam boilers. **[C]**
   A. deadweight
   B. chain
   C. spring-loaded pop-off
   D. none of the above

9. The total force on a safety valve 2½″ in diameter with a steam pressure of 15 psi is _____. **[C]**
   A. 19.5413
   B. 29.4525
   C. 73.63125
   D. 93.7512

10. The steam pressure gauge on the boiler is calibrated to read _____. **[B]**
    A. inches of vacuum
    B. pounds per square inch
    C. absolute pressure
    D. pressure below atmospheric pressure

11. Steam is prevented from entering the Bourdon tube of the pressure gauge by a(n) _____. **[D]**
    A. automatic nonreturn valve
    B. inspector's test cock
    C. os&y valve
    D. siphon

12. A _____ pressure gauge can read either vacuum or pressure. **[A]**
    A. compound
    B. duplex
    C. suction
    D. vacuum

13. Vacuum is pressure _____ pressure. **[C]**
    A. above gauge
    B. below gauge
    C. below atmospheric
    D. equal to gauge

14. Safety valves are designed to pop open and stay open until there is a(n) _____ psi drop in pressure. **[B]**
    A. 0 to 1
    B. 2 to 4
    C. 5 to 15
    D. over 15

___D___ 15. Safety valves are used on _____ boilers.
   A. firetube
   B. watertube
   C. cast iron
   D. all of the above

___A___ 16. The water column is located at the NOWL so the lowest visible part of the gauge glass is _____ above the highest heating surface.
   A. 2″ to 3″
   B. 4″ to 5″
   C. just
   D. never

___A___ 17. Blowback is the _____ in the boiler after the safety valve has opened.
   A. drop in pressure
   B. carryover
   C. increase of pressure
   D. chattering of a feedwater valve

___B___ 18. The boiler bottom blowdown line should discharge to a(n) _____.
   A. sewer
   B. blowdown tank
   C. atmospheric tank
   D. return tank

___B___ 19. If the desired cut-in pressure of the boiler is 6 psi and the desired cut-out pressure is 10 psi, the differential pressure setting must be _____ psi.
   A. 2
   B. 4
   C. 6
   D. 8

___D___ 20. Impurities that build up on the surface of the water in the boiler prevent _____ from breaking through the surface of the water.
   A. air
   B. oxygen
   C. CO_2
   D. steam

___C___ 21. To prevent air pressure from building up in the boiler when filling the boiler with water, the _____ must be open.
   A. safety valve
   B. main steam stop valve
   C. boiler vent
   D. manhole cover

____C____ 22. To prevent a vacuum from forming when taking the boiler off-line, the _____ must be open when pressure is still on the boiler.
   A. safety valve
   B. main steam stop valve
   C. boiler vent
   D. manhole cover

____C____ 23. The operating range of the boiler is controlled by the _____.
   A. aquastat
   B. vaporstat
   C. pressure control
   D. modulating pressure control

____D____ 24. The _____ regulates the high and low fire of the burner.
   A. aquastat
   B. vaporstat
   C. pressure control
   D. modulating pressure control

____B____ 25. The best time to blow down a boiler to remove sludge and sediment is when the boiler is under a _____ load.
   A. high
   B. light
   C. load twice the MAWP
   D. maximum

____C____ 26. The level of the water in the _____ indicates the water level in the boiler.
   A. condensate return tank
   B. try cocks
   C. gauge glass
   D. blowdown tank

____B____ 27. When blowing down a boiler, the quick-opening valve should always be opened _____ and closed _____.
   A. first; first
   B. first; last
   C. last; first
   D. last; last

____D____ 28. _____ added to boiler water change(s) scale-forming salts into a nonadhering sludge.
   A. Oxygen
   B. Minerals
   C. Slag
   D. Chemicals

29. A _____ provides access inside the water side of the boiler for inspection, sight, or cleaning. [C]
    A. boiler vent
    B. safety valve
    C. handhole
    D. siphon

30. A(n) _____ gauge is a pressure gauge that reads more pressure than is actually in the boiler. [D]
    A. broken
    B. slow
    C. uncalibrated
    D. fast

31. According to the ASME code, a try lever test on safety valves should be performed every _____ the boiler is in operation. [D]
    A. hour
    B. four hours
    C. seven days
    D. 30 days

32. The purpose of the safety valve is to prevent the pressure in the boiler from _____. [A]
    A. exceeding its MAWP
    B. dropping below its MAWP
    C. causing a boiler explosion
    D. relieving water pressure

33. The capacity of a safety valve is measured by the amount of steam that can be discharged per _____. [B]
    A. shift
    B. hour
    C. minute
    D. day

34. _____ is when a safety valve opens and closes rapidly. [C]
    A. Feathering
    B. Pressurizing
    C. Chattering
    D. Huddling

35. The safety valve on a low pressure boiler is designed to open when pressure in the boiler exceeds _____. [A]
    A. the safety valve setting
    B. the NOWL
    C. feedwater pump pressure
    D. the bottom blowdown discharge pressure

___A___ 36. After the total force of the steam has lifted the safety valve off its seat, steam enters the _____.
   A. huddling chamber
   B. combustion chamber
   C. steam holding tank
   D. main steam line

___C___ 37. _____ causes false water level readings in the gauge glass.
   A. Priming
   B. Carryover
   C. Foaming
   D. Blowing down

___A___ 38. The ASME Code requires fusible plugs on _____ boilers.
   A. coal-fired
   B. fuel oil-fired
   C. gas-fired
   D. all of the above

___B___ 39. On a pressure control, _____ pressure plus _____ pressure equals _____ pressure.
   A. differential; cut-out; cut-in
   B. cut-in; differential; cut-out
   C. cut-in; cut-out; differential
   D. cut-in; sequential; modulating

___A___ 40. A burner should always start up in _____ fire and shut down in _____ fire.
   A. low; low
   B. low; high
   C. high; low
   D. high; high

Safety Valve

___F___ 1. Valve disc
___E___ 2. Body
___A___ 3. Try lever pin
___B___ 4. Try lever
___H___ 5. Spindle
___C___ 6. Spring
___D___ 7. Valve seat
___G___ 8. Bonnet

Safety Valve Huddling Chamber

D	1. Valve seat
E	2. Huddling chamber
A	3. Spring
F	4. Valve spindle
C	5. Steam pressure
B	6. Valve disc

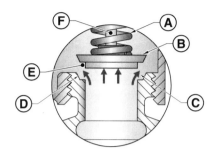

Steam Pressure Gauge Operation

C	1. Linkage
E	2. Siphon connection
D	3. Steam pressure
G	4. Face
F	5. Needle
A	6. Bourdon tube
B	7. Case

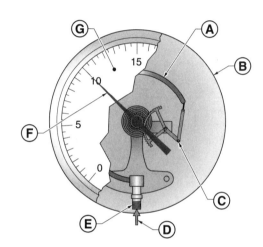

Water Column and Gauge Glass

C	1. Water column
D	2. Water column blowdown valve
A	3. Cross tees
I	4. Vent
G	5. Try cocks
F	6. Isolation valve
H	7. Gauge glass
B	8. NOWL
E	9. Gauge glass blowdown valve

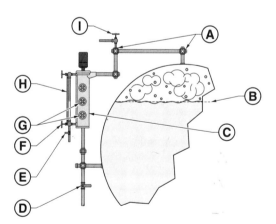

Bottom Blowdown Valve

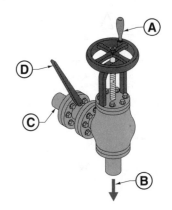

__D__ 1. Quick-opening valve
__A__ 2. Slow-opening valve
__B__ 3. To blowdown line
__C__ 4. From boiler

HRT Boiler

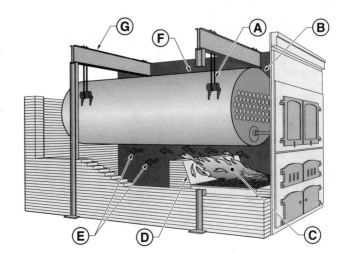

__C__ 1. Furnace
__E__ 2. Gases of combustion
__A__ 3. Sling
__B__ 4. Tube sheet
__D__ 5. Bridge wall
__F__ 6. Boiler
__G__ 7. Steel beam

Straight-Tube Watertube Boiler

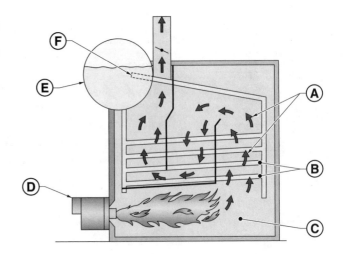

__E__ 1. Steam and water drum
__F__ 2. Internal feedwater line
__B__ 3. Tubes
__A__ 4. Gases of combustion
__C__ 5. Furnace
__D__ 6. Burner assembly

Feedwater System

Text Reference Pages 61 – 78

3

Name _____ Date _____

True-False

- (T) F 1. A high water level condition in the boiler could lead to water hammer.
- (T) F 2. A low water level condition in the boiler can result in damage to boiler heating surfaces.
- T (F) 3. The check valve [FEEDWATER STOP VALVE] on the feedwater line should be located closest to the shell of the boiler.
- (T) F 4. The check valve allows the flow of water in one direction only.
- (T) F 5. The check valve on the feedwater line commonly has a valve disc that swings to open and close the valve.
- (T) F 6. A condensate return tank collects condensate returned from heating units for use in the boiler.
- (T) F 7. A globe valve used as a feedwater stop valve must be installed so inlet pressure is applied from under the valve disc.
- (T) F 8. The stop valve on the feedwater line should be located closest to the boiler.
- (T) F 9. The vacuum pump is designed to discharge air and pump water.
- T (F) 10. Centrifugal [CONDENSATE] feedwater pumps are commonly driven by steam.
- T (F) 11. A check [GLOBE] valve is opened by turning the handle counterclockwise.
- (T) F 12. A low water fuel cutoff [OPENS AUTOMATICALLY] is located slightly below the NOWL.
- (T) F 13. The range of pressure on the vacuum switch on a vacuum pump is usually 2″ to 8″.
- T (F) 14. An evaporation test is performed by opening the low water fuel cutoff blowdown valve by hand. [SHUTTING OFF CITY MAKEUP FEEDER]
- (T) F 15. Any loss of water in the system must be made up by the makeup water feeder.
- (T) F 16. The boiler operator must be present during an evaporation test.
- T (F) 17. The makeup water feeder can be used as a feedwater regulator.

S ASME CODE SYMBOL STAMP | **POWER BOILER**

T F 18. Most makeup water contains some scale-forming salts.

T F 19. The function of the automatic makeup water feeder is to replace water that has been lost.

T F 20. The automatic city water makeup water feeder is located slightly below the NOWL.

Multiple Choice

__C__ 1. The water in the boiler is heated, turns to steam, and leaves the boiler through the ____ .
   A. feedwater line
   B. main header
   C. main steam line
   D. main branch line

__B__ 2. When steam releases heat in a heat exchanger, it turns to ____ .
   A. low pressure steam
   B. condensate
   C. makeup water
   D. exhaust steam

__D__ 3. A ____ pump returns condensate from the system back to the boiler.
   A. fuel oil
   B. return
   C. gear
   D. vacuum

__D__ 4. A(n) ____ valve allows the flow of water in one direction only.
   A. gate
   B. globe
   C. os&y
   D. check

__B__ 5. The feedwater ____ valve should be located as close to the shell of the boiler as practical.
   A. check
   B. stop
   C. nonreturn
   D. regulating

__B__ 6. A ____ after each heating unit allows ____ to pass through to the return line.
   A. steam trap; steam
   B. steam trap; condensate
   C. water trap; steam
   D. water trap; condensate

__D__ 7. A(n) _____ may be required on some boilers to remove oxygen and other gases.
   A. makeup water system
   B. pressure relief valve
   C. feedwater regulator
   D. feedwater heater

__C__ 8. The feedwater _____ valve opens and closes automatically.
   A. return
   B. bypass
   C. check
   D. equalizing

__D__ 9. The vacuum pump pumps water and discharges air to the _____.
   A. expansion tank
   B. compression tank
   C. return tank
   D. atmosphere

__C__ 10. The range of pressure on the vacuum switch is usually _____.
   A. 2 psi to 6 psi
   B. 6 psi to 12 psi
   C. 2″ to 8″
   D. 8″ to 12″

__B__ 11. The primary function of a low water fuel cutoff is to _____.
   A. remove sludge and sediment from the bottom of the boiler
   B. shut down the burner if the water level drops below the safe operating level
   C. stop the flow of city water supplied to the condensate return tank
   D. reduce water turbulence in the gauge glass

__B__ 12. Water added to the boiler to replace water lost due to leaks and blowing down is known as _____ water.
   A. extra
   B. makeup
   C. boiler
   D. feed

__B__ 13. Excessive use of cold city makeup water reduces overall boiler efficiency because the water must be _____ before use in the boiler.
   A. vented
   B. heated
   C. filtered
   D. reticulated

__B__ 14. An auxiliary low water fuel cutoff is installed _____.
   A. slightly above the NOWL
   B. slightly below the primary low water fuel cutoff
   C. at the highest point on the boiler
   D. on the feedwater line close to the boiler shell.

__C__ 15. The _____ shuts off the burner in the event of low water.
   A. low water alarm
   B. feedwater regulator
   C. low water fuel cutoff
   D. automatic low water feeder

__A__ 16. Water is supplied to the condensate return tank by the _____ pump.
   A. vacuum
   B. condensate
   C. feedwater
   D. return

__C__ 17. The feedwater regulator is located at the _____.
   A. boiler vent
   B. MAWP
   C. NOWL
   D. bottom blowdown line

__D__ 18. The _____ maintains a constant water level in the boiler.
   A. gauge glass
   B. water column
   C. blowdown tank
   D. feedwater regulator

__A__ 19. The low water fuel cutoff should be tested _____.
   A. daily
   B. monthly
   C. semiannually
   D. annually

__B__ 20. The burner should be _____ when the low water fuel cutoff is blown down.
   A. off
   B. firing
   C. tagged out
   D. tested

Feedwater System

_F_	1. Main steam header
_J_	2. Boiler
_A_	3. Steam trap
_I_	4. Stop valve
_O_	5. Heating unit
_K_	6. Water level
_B_	7. Condensate return line
_D_	8. Feedwater pump
_C_	9. Vent
_M_	10. Main steam stop valve
_H_	11. Check valve
_G_	12. Feedwater line
_E_	13. Condensate return tank
_N_	14. Riser
_L_	15. Main steam line

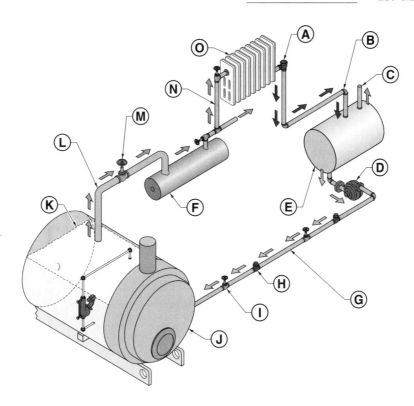

Testing Low Water Fuel Cutoff

_D_	1. Burner
_B_	2. Float at NOWL position
_A_	3. Mercury switch
_E_	4. Burner control
_C_	5. Float chamber

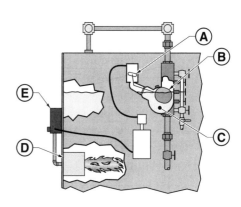

Makeup Water System

____J____ 1. Gauge glass
____D____ 2. Strainer
____E____ 3. Condensate return line
____A____ 4. Backflow preventer
____K____ 5. Low water fuel cutoff
____B____ 6. Makeup water supply
____G____ 7. Burner
____H____ 8. Automatic makeup water feeder
____C____ 9. Manual makeup water valve
____I____ 10. NOWL
____F____ 11. Blowdown valve

Feedwater Control

____C____ 1. Vent
____F____ 2. Feedwater pump
____L____ 3. Vacuum tank
____D____ 4. Condensate return tank
____J____ 5. NOWL
____K____ 6. Feedwater regulator and low water fuel cutoff
____I____ 7. Boiler
____G____ 8. Feedwater check valve
____E____ 9. Makeup water supply
____B____ 10. Vacuum pump
____H____ 11. Feedwater stop valve
____A____ 12. Condensate return line

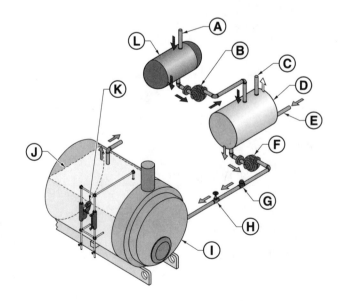

Feedwater System

Feedwater Line

_B_ 1. Check valve
_D_ 2. Main feedwater line
_E_ 3. Boiler
_A_ 4. From feedwater heater
_C_ 5. Stop valve

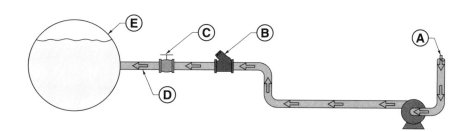

Vacuum Pump

_E_ 1. Condensate to boiler
_F_ 2. Vacuum tank
_B_ 3. Condensate from system
_G_ 4. Controls
_A_ 5. Vacuum switch
_D_ 6. Pump
_C_ 7. Float-controlled switches

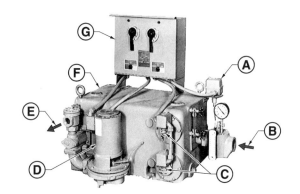

Centrifugal Water Pump

_D_ 1. Water enters inlet
_G_ 2. Housing
_B_ 3. Impeller rotation
_A_ 4. Electric motor
_F_ 5. Water thrown out
_C_ 6. Feedwater to boiler
_E_ 7. Impeller

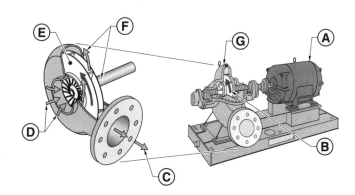

Low Water Fuel Cutoff Parts

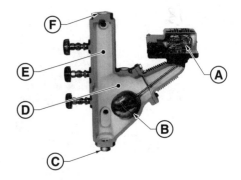

__B__ 1. Float
__E__ 2. Water column
__A__ 3. Control switches
__F__ 4. Steam connection
__C__ 5. Water connection
__D__ 6. Float chamber

Low Water Fuel Cutoff Operation

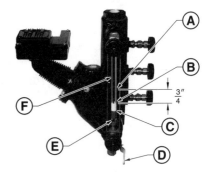

__B__ 1. Feedwater pump turns ON
__F__ 2. Gauge glass
__C__ 3. Burner shuts OFF
__D__ 4. Gauge glass blowdown valve
__A__ 5. Feedwater pump turns OFF
__E__ 6. Lowest visible point in gauge glass

Steam System

Text Reference Pages 79 – 92

4

Name_____ Date _____

True-False

- **(T)** F   1. An inverted bucket steam trap is a nonreturn trap.
- T **(F)**   2. A float thermostatic steam trap is a return trap.
- **(T)** F   3. A globe valve should never be used as a main steam stop valve.
- **(T)** F   4. Gate valves offer no restriction to flow when open.
- **(T)** F   5. A steam header is a distribution pipe that supplies steam to the branch lines.
- T **(F)**   6. Condensate in the steam lines helps to keep them clean.
- **(T)** F   7. Steam traps are located throughout the system after each device where steam is used.
- T **(F)**   8. Air trapped in the top of the bucket of an inverted bucket steam trap causes the steam trap to remain closed.
- **(T)** F   9. Return steam traps discharge condensate directly back to the boiler.
- T **(F)**   10. Return steam traps are commonly found in modern boiler plants.

Multiple Choice

___B___   1. Boiler main steam stop valves should be _____ valves.
- A. globe
- B. gate
- C. check
- D. safety

___C___   2. Gate valves should always be _____ or _____ closed.
- A. partially open; completely
- B. wide open; partly
- C. wide open; completely
- D. throttled; completely

___D___ 3. Steam header valves should be _____ valves.
   A. globe
   B. gate
   C. check
   D. os&y gate

___A___ 4. When open, an os&y valve offers _____ to the flow of steam.
   A. no resistance
   B. throttling action
   C. velocity
   D. full resistance

___C___ 5. Steam traps are _____ devices.
   A. manual
   B. electrical
   C. automatic
   D. semiautomatic

___D___ 6. _____ allow movement caused from the heating and cooling of steam lines.
   A. Lagging valves
   B. Bottom blowdown valves
   C. Header baffles
   D. Expansion bends

___A___ 7. Steam traps remove _____ and _____ from the steam lines.
   A. air; water
   B. air; oil
   C. water; oil
   D. air; steam

___D___ 8. Condensate in the steam lines can result in _____.
   A. greater boiler efficiency
   B. pure steam produced
   C. foaming
   D. water hammer

___C___ 9. Two types of steam traps are the _____ steam trap and the _____ steam trap.
   A. return; float
   B. nonreturn; thermostatic
   C. return; nonreturn
   D. nonreturn; float

__A__ 10. A(n) _____ steam trap is the most common steam trap used after a heating unit.
   A. thermostatic
   B. return
   C. inverted bucket
   D. thermodynamic

__D__ 11. Condensate from a nonreturn steam trap is pumped from the condensate return tank to the _____.
   A. return tank
   B. vacuum tank
   C. feedwater tank
   D. boiler

__B__ 12. Steam returning to the vacuum tank could cause the condensate pump to become _____.
   A. waterbound
   B. steambound
   C. waterlogged
   D. steamlogged

__A__ 13. Steam strainers should be located on the steam line _____.
   A. before the steam trap
   B. after the steam trap
   C. after the feedwater heater
   D. before the steam header

__D__ 14. A(n) _____ is a device used to test steam trap function by analyzing the sound waves emitted.
   A. infrared thermometer
   B. contact indicator
   C. sound flow indicator
   D. ultrasonic tester

__C__ 15. When an os&y valve is open, the stem is in the _____ position.
   A. floating
   B. locked
   C. up
   D. down

__B__ 16. A thermodynamic steam trap opens and closes by a(n) _____.
   A. float
   B. movable disc
   C. electric sensor
   D. flexible bellows

28 LOW PRESSURE BOILERS WORKBOOK

__A__ 17. In a float thermostatic trap, the float rises to discharge _____.
- A. condensate
- B. steam
- C. feedwater chemicals
- D. water and steam

__B__ 18. A steam trap that fails to open causes the heating unit to become _____.
- A. steambound
- B. waterlogged
- C. very hot
- D. contaminated with fuel oil

__C__ 19. _____ remove dirt and impurities that may cause the steam trap to malfunction.
- A. Vacuum pumps
- B. Globe valves
- C. Steam strainers
- D. Steam separators

__B__ 20. When a steam trap is functioning properly, there is a _____ difference in temperature between the steam trap inlet and outlet.
- A. 5°F to 10°F
- B. 10°F to 20°F
- C. 20°F to 30°F
- D. 40°F to 100°F

Main Steam Stop Valve

__D__	1. Main steam line
__B__	2. Condensate
__E__	3. Water level
__A__	4. Steam trap
__G__	5. Main steam stop valve
__I__	6. Heating unit
__F__	7. Steam
__H__	8. Riser
__C__	9. Main steam header

Low Pressure Boilers
Second Edition

Workbook Answer Key

© 2005 by American Technical Publishers, Inc.
All rights reserved

2 3 4 5 6 7 8 9 – 05 – 9 8 7 6 5 4 3 2

Printed in the United States of America

ISBN 0-8269-4352-7

AMERICAN TECHNICAL PUBLISHERS, INC.
HOMEWOOD, ILLINOIS 60430-4600

1 BOILER OPERATION PRINCIPLES

True-False — 1

1. T
2. F
3. T
4. T
5. T
6. T
7. F
8. T
9. F
10. T
11. T
12. T
13. F
14. T
15. T
16. F
17. T
18. F
19. T
20. F

Multiple Choice — 2

1. D
2. B
3. A
4. B
5. C
6. D
7. B
8. C
9. A
10. B
11. C
12. B
13. C
14. A
15. C
16. D
17. A
18. D
19. D
20. D
21. D
22. D
23. A
24. A
25. D

Boiler Systems — 5

1. A
2. C
3. B
4. E
5. D

Scotch Marine Boiler — 6

1. B
2. A
3. D
4. C

Steam Heating System — 6

1. D
2. G
3. E
4. I
5. C
6. H
7. A
8. B
9. F

2 BOILER FITTINGS

True-False — 7

1. T
2. F
3. F
4. F
5. F
6. T
7. F
8. T
9. T
10. T
11. F
12. T
13. T
14. T
15. F
16. F
17. T
18. F
19. T
20. F
21. F
22. F
23. T
24. F
25. T
26. F
27. T
28. F
29. F
30. T
31. T
32. T
33. T
34. F
35. T
36. T
37. T
38. F
39. F
40. T

Multiple Choice — 9

1. B
2. B
3. C
4. D
5. D
6. D
7. B
8. C
9. C
10. B
11. D
12. A
13. C
14. B
15. D
16. A
17. A
18. B
19. B
20. D
21. C
22. C
23. C
24. D
25. B
26. C
27. B
28. D
29. C
30. D
31. D
32. A
33. B
34. C
35. A
36. A
37. C
38. A
39. B
40. A

Safety Valve — 14

1. F
2. E
3. A
4. B
5. H
6. C
7. D
8. G

Safety Valve Huddling Chamber — 15

1. D
2. E
3. A
4. F
5. C
6. B

Steam Pressure Gauge Operation — 15

1. C
2. E
3. D

4. G
5. F
6. A
7. B

Water Column and Gauge Glass 15

1. C
2. D
3. A
4. I
5. G
6. F
7. H
8. B
9. E

Bottom Blowdown Valve 16

1. D
2. A
3. B
4. C

HRT Boiler 16

1. C
2. E
3. A
4. B
5. D
6. F
7. G

Straight-Tube Watertube Boiler 16

1. E
2. F
3. B
4. A
5. C
6. D

3 FEEDWATER SYSTEM

True-False 17

1. T
2. T
3. F
4. T
5. T
6. T
7. T
8. T
9. T
10. F
11. F
12. T
13. T
14. F
15. T
16. T
17. F
18. T
19. T
20. T

Multiple Choice 18

1. C
2. B
3. D
4. D
5. B
6. B
7. D
8. C
9. D
10. C
11. B
12. B
13. B
14. B
15. C
16. A
17. C
18. D
19. A
20. B

Feedwater System 21

1. F
2. J
3. A
4. I
5. O
6. K
7. B
8. D
9. C
10. M
11. H
12. G
13. E
14. N
15. L

Testing Low Water Fuel Cutoff 21

1. D
2. B
3. A
4. E
5. C

Makeup Water System 22

1. J
2. D
3. E
4. A
5. K
6. B
7. G
8. H
9. C
10. I
11. F

Feedwater Control 22

1. C
2. F
3. L
4. D
5. J
6. K
7. I
8. G
9. E
10. B
11. H
12. A

Feedwater Line 23

1. B
2. D
3. E
4. A
5. C

Vacuum Pump 23

1. E
2. F
3. B
4. G
5. A
6. D
7. C

Centrifugal Water Pump 23

1. D
2. G
3. B
4. A
5. F
6. C
7. E

Low Water Fuel Cutoff Parts 24

1. B
2. E
3. A
4. F
5. C
6. D

Low Water Fuel Cutoff Operation 24

1. B
2. F
3. C

4. D
5. A
6. E

4 STEAM SYSTEM

True-False 25
1. T
2. F
3. T
4. T
5. T
6. F
7. T
8. F
9. T
10. F

Multiple Choice 25
1. B
2. C
3. D
4. A
5. C
6. D
7. A
8. D
9. C
10. A
11. D
12. B
13. A
14. D
15. C
16. B
17. A
18. B
19. C
20. B

Main Steam Stop Valve 28
1. D
2. B
3. E
4. A
5. G
6. I
7. F
8. H
9. C

Steam Trap Operation 29
1. A
2. B
3. D
4. C
5. C
6. A
7. D

8. B
9. D
10. B
11. C
12. A
13. C
14. A
15. B
16. D

Steam Trap 29
1. C
2. A
3. D
4. B

Steam Trap Location 30
1. F
2. E
3. H
4. A
5. K
6. G
7. J
8. D
9. I
10. B
11. C

Steam Trap Testing Devices 31
1. C
2. D
3. B
4. A
5. E

Testing Steam Traps 32
1. B
2. C
3. A

5 FUEL SYSTEM

True-False 33
1. T
2. T
3. T
4. T
5. T
6. F
7. T
8. T
9. F
10. T
11. T
12. T

13. F
14. T
15. T
16. T
17. T
18. F
19. T
20. F
21. F
22. T
23. T
24. T
25. F
26. T
27. T
28. T
29. F
30. T
31. T
32. F
33. T
34. T
35. T
36. T
37. F
38. F
39. F
40. F
41. T
42. T
43. T
44. T
45. T
46. T
47. F
48. T

Multiple Choice 35
1. D
2. B
3. B
4. C
5. C
6. B
7. D
8. B
9. A
10. C
11. C
12. B
13. B
14. C
15. B
16. A
17. D
18. D
19. B
20. C
21. C
22. C
23. B
24. A
25. D
26. C
27. D

28. D
29. D
30. C
31. B
32. C
33. B
34. D
35. D
36. A
37. B
38. B
39. A
40. D
41. A
42. C
43. D
44. C
45. A
46. B
47. D
48. B
49. C
50. C
51. B
52. B
53. C
54. A
55. B
56. D
57. D
58. B
59. D
60. C

Fuel Oil Pump 43

1. B
2. E
3. A
4. F
5. C
6. D

Low Pressure Gas System 44

1. I
2. E
3. L
4. A
5. N
6. G
7. K
8. D
9. H
10. B
11. M
12. C
13. J
14. F

Screw-Feed Stoker 44

1. F
2. I
3. B

4. D
5. H
6. J
7. C
8. G
9. A
10. E

High Pressure Gas System 45

1. L
2. E
3. I
4. A
5. M
6. J
7. D
8. G
9. B
10. F
11. K
12. C
13. H
14. N

Ram-Feed Stoker 45

1. G
2. E
3. A
4. I
5. D
6. B
7. H
8. F
9. C

Fuel Oil Burner 46

1. D
2. A
3. C
4. B

Modulating Control 46

1. C
2. D
3. B
4. A

Flame Sensor 46

1. C
2. A
3. D
4. B

Fuel Oil System 47

1. G
2. L
3. A
4. E
5. J

6. C
7. N
8. H
9. F
10. K
11. B
12. M
13. I
14. D

Fuel Oil Strainer 48

1. G
2. F
3. A
4. E
5. I
6. B
7. H
8. D
9. C
10. J

6 DRAFT SYSTEM

True-False 49

1. T
2. F
3. F
4. T
5. T
6. T
7. F
8. T
9. T
10. F
11. T
12. T
13. F
14. T
15. F
16. T
17. T
18. T
19. T
20. T

Multiple Choice 50

1. B
2. D
3. D
4. D
5. B
6. B
7. A
8. B
9. D
10. A
11. D
12. B

Answers 5

13. C
14. B
15. A
16. B
17. A
18. C

Manometer 52

1. C
2. A
3. B

Natural Draft 53

1. E
2. C
3. A
4. F
5. B
6. D

Forced Draft 53

1. F
2. J
3. A
4. G
5. I
6. B
7. H
8. D
9. E
10. C

Draft Control 54

1. E
2. B
3. H
4. G
5. C
6. F
7. D
8. A

Stack 54

1. B
2. G
3. D
4. C
5. F
6. E
7. A
8. H
9. I

7 BOILER WATER TREATMENT

True-False 55

1. T
2. F
3. T
4. T
5. F
6. F
7. T
8. T
9. F
10. T
11. F
12. F
13. T
14. F
15. T
16. F
17. T
18. F
19. T
20. F
21. T
22. T
23. F
24. T
25. T

Multiple Choice 56

1. B
2. C
3. B
4. B
5. A
6. A
7. D
8. D
9. D
10. D
11. D
12. A
13. A
14. D
15. D
16. C
17. A

Sludge Removal 58

1. D
2. A
3. C
4. E
5. B

Priming and Carryover 59

1. D
2. A
3. B
4. E
5. C

Foaming 59

1. B
2. E
3. D
4. A
5. C

Water Treatment Log Readings 59

1. 8.9
2. 45
3. 16
4. 0
5. 2700

Bypass Feeder 60

1. E
2. B
3. D
4. A
5. C
6. F

8 BOILER OPERATION PROCEDURES

True-False 61

1. T
2. T
3. T
4. T
5. F
6. T
7. F
8. F
9. F
10. T
11. F
12. T
13. T
14. T
15. F
16. F
17. T
18. F
19. T
20. F
21. F
22. T
23. T
24. F
25. T
26. T
27. F
28. F
29. T
30. T
31. F
32. T
33. F
34. T
35. F
36. T
37. T
38. F
39. F

40. T
41. F
42. T
43. F
44. F
45. F

Multiple Choice 63
1. D
2. C
3. B
4. C
5. B
6. B
7. D
8. D
9. A
10. C
11. D
12. C
13. A
14. B
15. D
16. C
17. C
18. A
19. C
20. B
21. A
22. A
23. D
24. B
25. D
26. D
27. A
28. A
29. C
30. D
31. D
32. A
33. A
34. D
35. D
36. C
37. B
38. B
39. B
40. A

9 HOT WATER HEATING SYSTEMS

True-False 69
1. T
2. F
3. F
4. T
5. F
6. F
7. F
8. T
9. F
10. T
11. T
12. F
13. F
14. T
15. T
16. F
17. F
18. T
19. F
20. T
21. T
22. F
23. F
24. T
25. F
26. T
27. T
28. F
29. T
30. T

Multiple Choice 70
1. A
2. B
3. D
4. C
5. C
6. B
7. C
8. D
9. D
10. D
11. B
12. A
13. C
14. C
15. B
16. C
17. C
18. D
19. B
20. D
21. D
22. D
23. C
24. B
25. C

Natural Circulation Hot Water Heating System 74
1. G
2. A
3. H
4. C
5. F
6. E
7. B
8. D

Forced Circulation Hot Water Heating System 75
1. K
2. A
3. Q
4. H
5. T
6. E
7. O
8. S
9. B
10. P
11. F
12. L
13. D
14. I
15. M
16. G
17. C
18. N
19. J
20. R

Safety Relief Valve 76
1. D
2. F
3. B
4. E
5. A
6. G
7. C

Aquastat 76
1. B
2. G
3. I
4. D
5. F
6. H
7. C
8. E
9. A

10 COOLING SYSTEMS

True-False 77
1. T
2. T
3. T
4. F
5. T
6. F
7. F
8. T
9. F
10. T
11. T
12. F
13. T
14. T
15. T
16. F
17. T

18. F
19. T
20. F
21. T
22. T
23. F
24. T
25. F
26. T
27. F
28. F
29. T
30. T

Multiple Choice 78

1. B
2. C
3. D
4. C
5. D
6. D
7. C
8. D
9. A
10. C
11. B
12. B
13. A
14. D
15. A
16. B
17. D
18. B
19. C
20. C

Compression Refrigeration System 81

1. D
2. B
3. F
4. E
5. C
6. A

Liquid Receiver 82

1. E
2. B
3. A
4. D
5. C

Ammonia-Water Absorption Refrigeration System 82

1. G
2. B
3. C
4. F
5. D
6. A
7. E

Direct Cooling System 83

1. A
2. D
3. B
4. E
5. C

Cooling System Application – Pasteurization Process 83

1. D
2. B
3. C
4. G
5. F
6. A
7. E

Indirect Cooling System 84

1. B
2. D
3. F
4. A
5. E
6. G
7. C

11 BOILER OPERATION SAFETY

True-False 85

1. T
2. T
3. F
4. T
5. F
6. T
7. F
8. T
9. T
10. T
11. T
12. F
13. T
14. F
15. F
16. T
17. T
18. T

Multiple Choice 86

1. D
2. B
3. D
4. A
5. C
6. B
7. C

8. D
9. B
10. D
11. A
12. C
13. D
14. A
15. D

Hazardous Material Container Labeling – RTK Labeling 88

1. D
2. F
3. A
4. E
5. B
6. C

12 TESTING

Test 1
True-False 89

1. T
2. F
3. T
4. F
5. F
6. T
7. T
8. F
9. T
10. T
11. T

Multiple Choice 90

1. A
2. B
3. B
4. B
5. D
6. C
7. C
8. A
9. A
10. C
11. D
12. C
13. A
14. B
15. B
16. A
17. C
18. D
19. C
20. C
21. A
22. B

23. D
24. C
25. A
26. D
27. B
28. B
29. A
30. C
31. D
32. D
33. B
34. B
35. B
36. C
37. C
38. B
39. A
40. D
41. D
42. A
43. B
44. B
45. D

Test 2
True-False 97

1. T
2. F
3. F
4. T
5. T
6. F
7. T
8. T
9. F
10. T

Multiple Choice 97

1. C
2. D
3. B
4. C
5. C
6. A
7. B
8. B
9. A
10. A
11. A
12. C
13. D
14. D
15. C
16. B
17. D
18. B
19. A
20. A
21. A
22. B
23. A
24. C
25. C
26. D

27. A
28. D
29. B
30. D
31. A
32. B
33. B
34. D
35. B
36. C
37. D
38. A
39. D
40. B
41. B
42. C
43. A
44. C
45. C

Test 3
True-False 105

1. F
2. T
3. T
4. T
5. F
6. T
7. T
8. F
9. T
10. T
11. T

Multiple Choice 105

1. D
2. A
3. D
4. B
5. D
6. C
7. A
8. A
9. B
10. B
11. B
12. C
13. B
14. C
15. A
16. D
17. C
18. A
19. A
20. A
21. D
22. B
23. B
24. D
25. A
26. A
27. A
28. B
29. B

30. B
31. D
32. B
33. D
34. A
35. D
36. A
37. A
38. A
39. A
40. A
41. C

Test 4
True-False 111

1. T
2. F
3. F
4. T
5. T
6. F
7. T
8. F
9. F
10. T
11. F

Multiple Choice 111

1. B
2. A
3. C
4. A
5. A
6. B
7. C
8. D
9. A
10. C
11. D
12. C
13. B
14. C
15. A
16. D
17. C
18. B
19. C
20. D
21. B
22. D
23. D
24. C
25. C
26. C
27. C
28. B
29. A
30. C
31. D
32. B
33. D
34. C
35. C
36. B

37. C
38. A
39. C
40. B
41. A

Test 5
True-False   117

1. T
2. F
3. T
4. F
5. T
6. T
7. T
8. T
9. T
10. F
11. T
12. F
13. F

Multiple Choice   117

1. D
2. A
3. B
4. C
5. B
6. C
7. B
8. C
9. C
10. A
11. D
12. C
13. C
14. A
15. B
16. A
17. C
18. B
19. B
20. C
21. C
22. C
23. C
24. C
25. C
26. C
27. C
28. C
29. D
30. B
31. A
32. C
33. C
34. C
35. B
36. D
37. A
38. C
39. D
40. A

Test 6
True-False   123

1. T
2. T
3. F
4. F
5. T
6. T
7. T
8. F
9. T
10. F
11. F
12. T
13. F
14. F
15. T

Multiple Choice   123

1. B
2. C
3. C
4. C
5. C
6. B
7. B
8. D
9. B
10. C
11. D
12. D
13. A
14. B
15. D
16. A
17. D
18. C
19. D
20. C
21. A
22. D
23. C
24. A
25. C
26. B
27. C
28. D
29. A
30. D
31. A
32. C
33. B
34. A
35. B
36. C
37. B
38. C
39. D
40. A
41. B

Test 7
Essay   131

1. Report to work 10 to 15 minutes early. Check the water level on all the boilers on the line by blowing down the gauge glass, water column, and low water fuel cutoff. *NOTE*: (1) The burners should be firing when the low water fuel cutoff is blown down. This not only removes any sludge and sediment from the lines, but also shuts off the burner to ensure its proper operation. (2) When blowing down the gauge glass, the water should leave the gauge glass quickly and return quickly when the gauge glass blowdown valve is closed. This indicates that the lines are clean.

2. Steam coming out of the bottom try cock is an indication of low water condition in the boiler. The burner must be secured, the boiler taken off-line and cooled slowly, and the chief engineer and/or boiler inspector must be notified so the boiler can be inspected for any overheating of the heating surface. *NOTE*: Never add water to a boiler that has overheated before checking for possible damage.

3. Water coming out of the top try cock is an indication of a high water level condition. The boiler must be given a bottom blowdown to prevent carryover and possible water hammer. The operator should then determine the cause of the high water level condition.

4. The different ways of getting water to the boiler are the vacuum pump, feedwater pump, automatic city water makeup feeder, and hand-operated city water makeup valve.

5. The pressure control is sensitive to pressure and controls the boiler operating range. It starts and stops the burner on steam pressure demand.

6. Safety valves are tested by hand or by pressure. The ASME code recommends that safety valves be tested at least once a month by raising the boiler pressure and causing the safety valve(s) to pop open. When testing a safety valve by hand there should be at least 5 psi of pressure before lifting the test lever. The hand test lever is lifted to fully open the safety valve and is released to allow the safety valve to snap shut. *NOTE*: Check with local inspectors or the mechanical inspection bureau regarding the proper testing of safety valves.

7. According to Section VI of the ASME code, safety valves on low pressure boilers should be tested by hand once every 30 days the boiler is in operation.

The ASME code also recommends the testing of safety valves under pressure once a year, preferably before the start of the heating season. In addition to the ASME code, safety valves should be tested according to recommendations of the local boiler inspector.

8. The function of the low water fuel cutoff is to shut the burner off to protect the boiler heating surface from becoming overheated due to low water in the boiler.

9. The two methods of testing the low water fuel cutoff are by hand or by an evaporation test. When testing the burner by hand, the blowdown valve on the low water fuel cutoff is opened and steam and water rushing out cause the float in the low water fuel cutoff to drop, resulting in the burner shutting off. When performing an evaporation test, the vacuum pump, feedwater pump (if the boiler has one), and the automatic city water makeup feeder must be secured. The water level will drop in the boiler and the float in the low water fuel cutoff will fall slowly, shutting the burner off. *NOTE*: When the burner shuts off, there must always be water still visible in the gauge glass.

10. The maximum allowable working pressure (MAWP) for a low pressure steam boiler is 15 psi. A safety valve must pop (open) whenever pressure exceeds the safety valve setting to prevent the boiler from exceeding its MAWP. Anytime a safety valve pops, the boiler operator must determine the cause and take corrective action.

11. The safety valves on low pressure steam boilers are set to pop open at 15 psi.

12. Purging a furnace is the part of the firing cycle when the burner motor fan blows air into the furnace to remove any mixture of gases or fumes of a combustible nature that might cause a furnace explosion. The furnace is always purged before the burner lights off (prepurge) and starts again after the burner shuts off (postpurge).

13. The most important valve on the boiler is the safety valve. The safety valve prevents the boiler from exceeding its MAWP. If the safety valve exceeds its MAWP, a boiler explosion could result.

14. The boiler should be given a bottom blowdown whenever tests of the boiler water indicate high chemical concentration or high total dissolved solids. The boiler should also be blown down when a high water level develops.

15. The flame scanner can be tested in two ways. One method is to remove the scanner from its sighting tube and cover it with your hand, which should shut off the burner. The second method is to secure the fuel going to the burner to simulate a fuel failure, which should cause the scanner to shut the burner off on safety lockout.

16. There are two methods of determining the water level in the boiler. One method is to blow down the gauge glass and watch the action of the water leaving and returning to the glass. The second method is to use the three try cocks and observe what comes out of the top try cock, middle try cock, and bottom try cock when they are opened. With an NOWL the top try cock should discharge steam, the middle try cock should discharge a mixture of water and steam, and the bottom try cock should discharge water.

17. Smoke is the result of incomplete combustion. For example, in fuel oil-burning plants, smoke could be caused by cold fuel oil, poor atomization, insufficient primary and secondary air, a dirty burner, or fuel oil impinging on the brickwork.

18. The two types of draft used in boilers are natural draft and mechanical draft. Mechanical draft can be further classified as forced draft, induced draft, or combination forced and induced draft.

19. The check valve on the feedwater line prevents water from leaving the boiler when the vacuum pump or boiler feedwater pump stops. A check valve controls the flow of feedwater in one direction only.

20. The stop valve is located on the feedwater line as close to the shell of the boiler as practical between the boiler and the check valve. The stop valve is required in the event the check valve malfunctions (sticks open or closed). The stop valve could then be secured (closed) and the check valve repaired without having to dump the boiler.

21. The safety valve popping and steam pressure of 30 psi would indicate malfunction of the safety valve(s). The procedure to follow would be to: (1) Secure the burner. (2) Allow steam pressure to drop to approximately 15 psi. (3) Use the safety valve test lever to test the safety valve(s) by hand. (4) If the safety valve(s) pop open and reset at the proper blowback, raise the steam pressure to see if the safety valve(s) popped by pressure. (5) If the safety valve(s) popped by pressure, determine why the pressure control did not function properly. *NOTE*: The boiler in question should not be left unattended. A new boiler should be warmed up and cut in on the line if there is one available. If this is the only boiler in the plant, it must be attended 24 hours a day until it can be taken off-line and given a complete safety check.

22. Perfect combustion is the burning of all the fuel using the theoretical amount of air. Perfect combustion can only be achieved in a laboratory setting. Complete combustion is the burning of all the fuel using the minimum amount of excess air. Incomplete combustion is when all the fuel is not burned, resulting in soot and smoke.

23. Furnace explosions are caused by a buildup of highly combustible gases or fumes that ignite when exposed to a spark, pilot flame, or radiant brickwork. In addition, furnace explosions can result from improper purging of the furnace after a flame failure, a leaking fuel valve allowing fuel to enter the furnace, or the operator bypassing safety interlocks in order to light the burner quicker after a fuel interruption.

24. Before a boiler can be inspected, it must be taken off-line. Once a boiler is off-line, the following safety checks must be made: (1) The main steam stop valve(s) (some boilers have two) must be closed and tagged out. To tag out a valve means to mark it so it will not be opened by mistake. Some plants attach a sign to the valve wheel that reads "Danger! Personnel in boiler–do not open." (2) The boiler vent or top try cock should be checked to see that it is open. This ensures there is no vacuum inside. (3) The feedwater line to the boiler must be closed and tagged out. If there is an automatic city water makeup valve, it must be secured also.

25. A gauge glass should be blown down once a shift or whenever the operator questions the level of water in the boiler.

26. In a forced circulation system, the system is closed to the atmosphere and water is forced (circulated) through the heating system using circulating pumps. In a natural circulation system, the system is open to the atmosphere. It works on the principle that water when heated increases in volume (expands) and becomes lighter. The warmer water rises and cold, heavier water flows back into the bottom of the boiler.

27. In order to do a hydrostatic test on a boiler, the boiler must be completely filled with water. To perform a hydrostatic test, the following operations must be carried out: (1) If the water column has a whistle valve, it must be removed and plugged. (2) The main steam stop valve must be closed. (3) The safety

valve(s) must be removed and blank flanges installed, or the safety valves must be gagged. (A gag is a clamp that prevents the valve from popping open without damaging the valve.) (4) The boiler vent must remain open until water comes out and then is closed. (5) Pressure on the boiler is brought up to 1½ times the MAWP. (The pressure must be under control so that it does not exceed this pressure by more than 10 pounds.)

28. NOWL stands for the normal operating water level. In a steam boiler the NOWL is approximately one-third to one-half a gauge glass.

29. MAWP stands for the maximum allowable working pressure of a boiler. The MAWP of a given boiler is determined by the ASME code.

30. Carryover can be caused by (1) carrying too high a water level, (2) high boiler water surface tension, and (3) opening the boiler main steam stop valve too quickly.

31. Natural draft is created by the height of the chimney and the difference in temperature of the gases of combustion in the chimney and the air outside the chimney.

32. A feedwater pump becomes steambound when the temperature of the water in the open feedwater heater gets too hot and water in the suction line going to the feedwater pump flashes into steam.

33. A fast gauge is a pressure gauge that reads more steam pressure than is actually present in the boiler. A slow gauge is a pressure gauge that reads less steam pressure than is actually present in the boiler.

34. Foaming is caused by impurities in the boiler water. A surface blowdown is used to remove impurities in the boiler water that cause foaming.

35. A gate valve, when open offers no restriction to the flow of the material passing through it. A gate valve must always be fully open or fully closed. A globe valve has material flowing under a valve seat. The globe valve is used for throttling (varying the rate of flow) service by partly opening or closing the valve.

36. In a water tube boiler, water passes through tubes that are surrounded by gases of combustion. In a firetube boiler, gases of combustion pass through tubes that are surrounded by water.

37. The water column must be located at the NOWL so that the lowest visible part of the gauge glass is 2" to 3" above the highest heating surface. The top line to the water column is connected to the highest part of the steam side of the boiler. The bottom line to the water column is connected approximately 6" below the center line of a firetube boiler.

38. The best time to give a boiler a bottom blowdown is when the boiler is at its lightest load. This allows sludge and sediment to fall to the bottom.

39. The function of a boiler vent is to: (1) vent air from the boiler when filling it with water, (2) vent air from the boiler when warming up prior to cutting in on the line, and (3) prevent a vacuum from forming in the boiler when taking the boiler off-line.

40. The high and low fire of the burner are controlled by the modulating pressure control through the modulating motor.

41. The os&y valve is open when its stem is in the up position.

42. The purpose of the vacuum pump is to help return the condensate back to the vacuum tank. At the vacuum tank, air is vented to the atmosphere, and water is pumped directly back to the boiler or a condensate return tank.

43. The automatic city water makeup feeder is a safety device that ensures that the boiler will have water if the vacuum pump fails or condensate is lost in the system. For maximum protection and safety, the automatic city water makeup feeder should be blown down daily to ensure its proper operation in an emergency.

44. The steam traps most commonly used are the nonreturn type. This type includes thermostatic, inverted bucket, and float thermostatic.

45. The two methods most commonly used to test the proper function of steam traps are strap-on thermometers and temperature-indicating crayons.

46. Fuel oil strainers should be cleaned daily and whenever unusually high suction readings occur when fuel oil is at its proper temperature.

47. Two types of fuel oil burners commonly used in low pressure plants are rotary cup burners and air atomizing burners.

48. The advantage of using a combination fuel oil/gas burner is plant flexibility; the operator can burn the fuel that is cheapest, change from one fuel to another if problems occur, and keep the plant operating if there is a shortage of a fuel.

49. Air used in the combustion process is broadly classified as primary, secondary, and excess air. Primary air is air that controls the combustion rate which determines the amount of fuel burned. Secondary air is air that controls combustion efficiency by controlling how completely the fuel is burned. Excess air is air supplied to the burner above the theoretical amount required to burn the fuel.

50. A flame safeguard system is burner control equipment that monitors the burner start-up sequence and the main flame during normal operation. The flame safeguard programmer sequences burner function in a set order including prepurge, ignition trials, pilot flame-establishing period, main burner flame-establishing period, run period, and postpurge.

51. Hydrostatic pressure is pressure caused by the weight of water (approximately .433 psi per vertical foot). Hydrostatic pressure can affect a steam pressure gauge reading depending on the distance from the steam boiler connection. Hydrostatic pressure must also be considered when sizing circulating pumps for hot water heating systems in multiple-story buildings.

52. Three types of heat transfer that occur in a boiler are conduction, convection, and radiation. Conduction is heat transfer that occurs when molecules in a material are heated and the heat is passed from molecule to molecule through the material. Convection is heat transfer that occurs when currents circulate between warm and cool regions of a fluid. Radiation is heat transfer that occurs as radiant energy (electromagnetic waves) without a material carrier.

53. A backflow preventer is a boiler accessory that prevents the flow of boiler water back to the city water supply. A backflow preventer is located on the makeup water supply line and functions like a check valve to allow water flow in one direction only.

54. Expansion bends are sometimes required on steam lines to allow for the expansion and contraction which occurs from temperature variation. Without proper expansion bends, damage to piping may occur from movement in steam lines.

55. An evaporation test is a test performed to ensure proper operation of a low water fuel cutoff. During an evaporation test, a low water condition is simulated by securing all water fed to the boiler. The boiler water level is then gradually lowered and the low water fuel cutoff should shut off the burner. Because a low water condition is created, an evaporation test must only be performed under the close supervision of authorized personnel.

56. A boiler room log is a record of information pertaining to the operation of a boiler

during a given period of time. Maintaining a boiler room log allows evaluation of boiler performance history such as steam pressure generated, water treatment, water levels, fuel consumption, and condensate returns. Boiler room log information can be used to determine the cause of a malfunction or help prevent a future problem.

57. An aquastat is a hot water boiler fitting that senses and controls water temperature by starting or stopping the burner. An aquastat functions like a switch using a boiler water temperature setpoint to turn the burner ON or OFF.

58. A compression refrigeration system uses a compressor, refrigerant, and pressure control devices. A compression refrigeration system is commonly divided into the high-pressure side and the low-pressure side. In the high-pressure side, refrigerant is changed from a gas to a liquid. A compressor compresses refrigerant vapor, and heat is released in the condenser. In the low-pressure side, the refrigerant is changed from a liquid to a gas. An expansion valve allows a drop in pressure, and the liquid refrigerant passes to the evaporator where it changes to a low-pressure gas and absorbs heat.

59. A confined space is a space large enough and so configured that a worker can physically enter and perform assigned work, has limited or restricted means for entry or exit, and is not designed for continuous occupancy. To ensure worker safety, each facility must have specific confined space procedures to comply with OSHA standards in the Code of Federal Regulations. Depending on the confined space and task, a confined space permit may be required. A confined space permit is a document that details requirements such as lock out, personal protective equipment, and ventilation requirements. An entry permit must be posted at confined space entrances before entering a permit-required confined space.

60. A boiler operator license is obtained by meeting the requirements established by the specific jurisdiction in which the boiler is operated. These requirements include passing a boiler operator licensing examination. The licensing jurisdiction can be a state, county, city, or other authority which has specific grades of licensing examinations for different sizes of boilers and related equipment. Some jurisdictions require experience in the field prior to taking a licensing examination. For specific requirements, the licensing jurisdiction should be contacted directly.

ISBN 0-8269-4352-7

9 780826 943521

Steam Trap Operation

____A____ 1. Condensate discharge outlet
____B____ 2. Inverted bucket
____D____ 3. Discharge valve
____C____ 4. Steam and condensate inlet

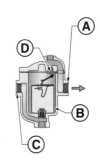

____C____ 5. Steam and condensate inlet
____A____ 6. Movable disc closed position
____D____ 7. Trapped steam in control chamber
____B____ 8. Condensate discharge outlet

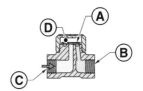

____D____ 9. Steam and condensate inlet
____B____ 10. Discharge valve
____C____ 11. Condensate discharge outlet
____A____ 12. Bellows

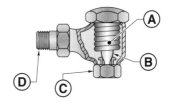

____C____ 13. Discharge valve
____A____ 14. Steam and condensate inlet
____B____ 15. Condensate discharge outlet
____D____ 16. Ball float

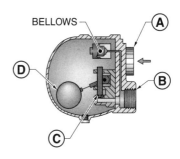

Steam Trap

____C____ 1. Thermodynamic
____A____ 2. Inverted bucket
____D____ 3. Thermostatic
____B____ 4. Float thermostatic

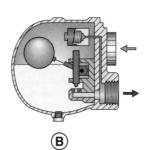 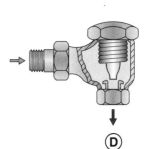

A　　　　　　B　　　　　　C　　　　　　D

Steam Trap Location

__F__ 1. Feedwater pump

__E__ 2. Check valves

__H__ 3. Main steam line

__A__ 4. Steam trap

__K__ 5. Heating unit

__G__ 6. Stop valves

__J__ 7. Main steam stop valve

__D__ 8. Condensate return tank

__I__ 9. Strainer

__B__ 10. Main steam header

__C__ 11. Condensate return line

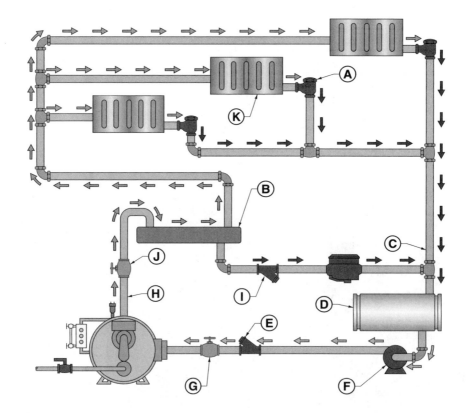

Steam Trap Testing Devices

C   1. Contact thermometer
D   2. Temperature-indicating crayon
B   3. Flow indicator
A   4. Ultrasonic tester
E   5. Infrared thermometer

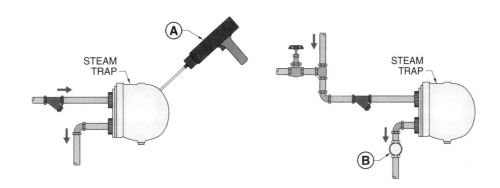

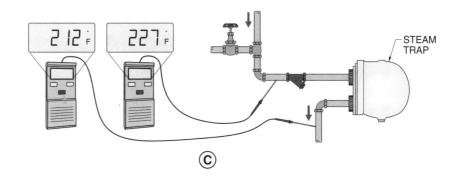

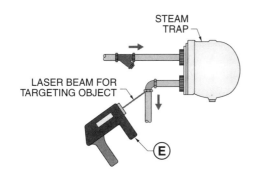

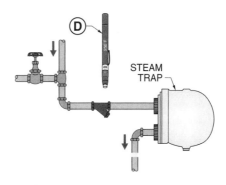

Testing Steam Traps

__B__ 1. Steam trap functioning normally

__C__ 2. Steam trap malfunction

__A__ 3. Strainer clogged

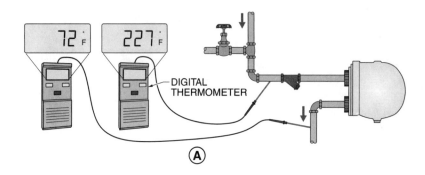

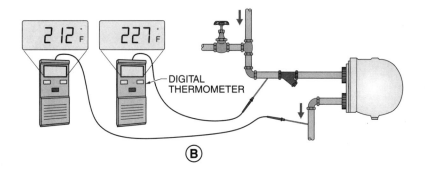

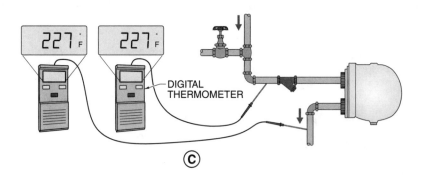

Fuel System

Text Reference Pages 93 – 138

5

Name _____ Date _____

True-False

T	F	1. Fuel oil and gas are the most commonly used fuels in low pressure boilers.
T	F	2. The fire point temperature is higher than the flash point temperature.
T	F	3. A fuel oil relief valve is located between the fuel oil pump and the fuel oil discharge valve.
T	F	4. The fuel oil pump should be started with its discharge valve open.
T	F	5. Cold fuel oil in the tank will give a high suction reading.
T	**F**	6. A dirty strainer results in low suction readings. *(high)*
T	F	7. Starting the fuel oil pump with its discharge valve closed causes the relief valve to open.
T	F	8. Fuel oil burners commonly use atomizing burners.
T	**F**	9. Rotary cup burners can only burn No. 2 fuel oil. *(4,5,6)*
T	F	10. Rotary cup burners atomize fuel oil using a spinning cup and high velocity air.
T	F	11. Air used to atomize fuel oil is primary air.
T	F	12. Secondary air is needed to burn the fuel oil efficiently.
T	**F**	13. In an air atomizing burner, steam and fuel are discharged to the burner for combustion. *(AIR)*
T	F	14. The air atomizing burner uses compressed air to mix with the fuel oil to achieve a high degree of atomization.
T	F	15. Complete combustion is achieved in a gas burner by supplying the proper mixture of air and gas to the furnace.
T	F	16. In a rotary cup burner, the solenoid valve controls the flow of fuel through the fuel tube.
T	F	17. Fuel oil to be removed from the burner nozzle is purged with air.

ASME CODE SYMBOL STAMP   PRESSURE PIPING

34 LOW PRESSURE BOILERS WORKBOOK

T	F	
T	**F**	18. A metal scraper is used to clean deposits on a fuel oil burner nozzle.
T	F	19. In a low pressure gas system, before the manual reset valve can be opened, the pilot must be lit.
T	**F**	20. The manual reset valve in the low pressure gas system closes if the boiler has a high water level. ~~high~~ LOW
T	**F**	21. A fuel oil heater is required when using No. 2 fuel oil. 456
T	F	22. Fuel oil is a liquid fossil fuel.
T	F	23. Plant flexibility is increased with the use of a combination gas/fuel oil burner.
T	F	24. Anthracite coal is hard coal.
T	**F**	25. Hard coal has a high volatile content. SOFT
T	F	26. Bituminous coal is soft coal.
T	F	27. Boilers burning soft coal need large furnace volumes to complete combustion.
T	F	28. Gases of combustion that contact boiler heating surfaces before combustion is completed cause soot and smoke.
T	**F**	29. Grates are only needed in hand-fired coal boilers to support the coal.
T	F	30. Smoke is a sign of incomplete combustion.
T	F	31. Smoke is less of a problem when burning hard coal as compared to soft coal.
T	**F**	32. Stokers were developed to reduce the amount of coal fed to the furnace. EFFICIENT/MORE COAL
T	F	33. Stoker firing allowed the design of larger coal-fired boilers.
T	F	34. The type of fuel used determines the fuel system accessories required.
T	F	35. Purging the lines and nozzles of an air atomizing burner at the end of its firing cycle keeps lines and nozzles clean for the next starting cycle.
T	F	36. Natural gas is a colorless and odorless fossil fuel.
T	**F**	37. Propane is lighter than natural gas and requires special handling. HEAVIER
T	**F**	38. The viscosity of fuel oil is the temperature at which it will burn continually when exposed to an open flame. INTERNAL RESISTANCE TO FLOW
T	**F**	39. The flash point of fuel oil is the temperature at which it will resist burning.
T	**F**	40. It is best to burn fuel oil with a low flash point. DANGEROUS
T	F	41. Pour point is the lowest temperature at which fuel oil will flow as a liquid.
T	F	42. A therm is the quantity of gas required to produce 100,000 Btu.
T	F	43. The temperature of the fuel in the fuel oil tank should be maintained at approximately 100°F. —120°F
T	F	44. Fuel oil in the storage tank must be kept below its flash point temperature.

(T) F   45. The temperature of the fuel oil in the storage tank must be kept above the recommended pour point.

(T) F   46. No. 6 fuel oil must be heated to the proper temperature in order to burn.

T   (F) 47. The flame scanner proves the main flame only. *PILOT*

(T) F   48. The furnace must be purged after any flame failure.

Multiple Choice

_D_ 1. When burning No. 6 fuel oil, the fuel oil strainers should be cleaned at least once every _____ hours.
  A. 8
  B. 10
  C. 12
  D. 24

_B_ 2. When cleaning a fuel oil strainer, the _____ must be carefully replaced to prevent air from entering the system.
  A. plug valve
  B. gasket
  C. hand screw
  D. basket

_B_ 3. The _____ pump draws fuel oil from the fuel oil tank.
  A. transfer
  B. fuel oil
  C. condensate
  D. circulating

_C_ 4. The _____ valve protects the fuel lines and pump from excessive pressure.
  A. safety
  B. bypass
  C. relief
  D. stop

_C_ 5. A high vacuum on the fuel oil suction gauge normally indicates _____.
  A. low viscosity or clogged vent
  B. a closed discharge valve or hot fuel oil
  C. cold fuel oil or a dirty strainer
  D. water in the fuel oil or a worn pump

_B_ 6. Fuel oil burners are designed to provide fuel oil to the furnace in a _____.
  A. steady stream
  B. fine spray
  C. half spray, half stream
  D. none of the above

D _____ 7. The rotary cup burner uses _____ to atomize the fuel oil.
   A. high temperature steam and pressure
   B. high temperature air and pressure
   C. steam and No. 6 fuel oil
   D. a spinning cup and high velocity air

B _____ 8. In a low pressure gas burner, gas is mixed with air in the _____.
   A. burner register
   B. mixing chamber before the burner register
   C. combustion chamber
   D. boiler furnace

A _____ 9. In a high pressure gas burner, the gas mixes with the air on the inside of the _____.
   A. burner register
   B. mixing chamber
   C. combustion chamber
   D. boiler furnace

C _____ 10. A _____ is a narrowed portion of a tube.
   A. solenoid
   B. butterfly valve
   C. venturi
   D. pilot regulator

C _____ 11. On a low pressure gas system, the manual reset cannot be opened until the _____.
   A. vaporstat proves pressure
   B. boiler is vented
   C. pilot is lighted
   D. all of the above

B _____ 12. Before repair work is attempted on a gas-fired boiler, the _____.
   A. insurance inspector must be notified
   B. main gas cock must be secured
   C. state inspector must be notified
   D. main solenoid valve should be secured

B _____ 13. In a high pressure gas system, the plant pressure _____ valve reduces gas pressure to line pressure used in the system.
   A. butterfly
   B. regulating
   C. safety
   D. pilot

14. _____ coal is coal that is ground to a fine powder.
    A. Anthracite
    B. Bituminous
    C. Pulverized
    D. none of the above

C

15. A _____ is a mechanical device for feeding coal consistently to the burner.
    A. grate burner
    B. stoker
    C. hopper shoveler
    D. all of the above

B

16. A(n) _____ is an air pressure-activated switch that closes after proving sufficient pressure of combustion air from the forced draft fan.
    A. air proving switch
    B. pilot modulating relay
    C. combustion flowmeter
    D. pressure switch control

A

17. A combination gas/fuel oil burner allows the operator to switch fuels _____.
    A. for economy
    B. if there is a shortage of a fuel
    C. if there is a failure in the fuel system being used
    D. all of the above

D

18. Hard coal is _____.
    A. bituminous coal with a high carbon content
    B. bituminous coal with a high volatile content
    C. anthracite coal with a high volatile content
    D. anthracite coal with a high carbon content

D

19. Soft coal is _____.
    A. bituminous coal with a high carbon content
    B. bituminous coal with a high volatile content
    C. anthracite coal with a high volatile content
    D. anthracite coal with a high carbon content

B

20. A(n) _____ sensor senses light frequencies that are higher than those visible to the eye.
    A. photocell
    B. infrared
    C. ultraviolet
    D. all of the above

C

38 LOW PRESSURE BOILERS WORKBOOK

C____ 21. A(n) _____ is a safety device that senses if the pilot light and/or main flame are lit.
  A. low water fuel cutoff
  B. annunciator
  C. flame scanner
  D. none of the above

C____ 22. The _____ of fuel oil is the lowest temperature at which it will flow as a liquid.
  A. fire point
  B. flash point
  C. pour point
  D. viscosity

B____ 23. The _____ of fuel oil is the temperature at which fuel oil gives off vapor that flashes when exposed to an open flame.
  A. fire point
  B. flash point
  C. pour point
  D. viscosity

A____ 24. The _____ is the temperature at which fuel oil will burn continuously when exposed to an open flame.
  A. fire point
  B. flash point
  C. pour point
  D. viscosity

D____ 25. The internal resistance of fuel oil to flow is _____.
  A. fire point
  B. flash point
  C. pour point
  D. viscosity

C____ 26. In order to lower the viscosity of fuel oil, it is necessary to _____.
  A. lower its temperature
  B. lower its pour point
  C. increase its temperature
  D. increase its pour point

D____ 27. A leak on the fuel oil suction line between the tank and the suction side of the fuel oil pump would result in _____.
  A. the suction gauge pulsating
  B. air entering the suction line
  C. pulsating of the fire in the boiler
  D. all of the above

Fuel System

D 28. Stokers were developed to _____.
- A. increase the efficiency of burning coal
- B. keep furnace temperatures constant to protect brickwork
- C. allow for development of larger coal-fired boilers
- D. all of the above

D 29. Foreign matter in the coal hopper of the screw-feed stoker is best removed by _____.
- A. emptying the coal hopper
- B. reversing the stoker
- C. forcing it through with a heavy shear pin
- D. using the cutoff gate at the bottom of the hopper

C 30. A _____ is used to prevent damage to the transmission of a screw-feed stoker in the event of an obstruction clogging the feed screw.
- A. slip clutch
- B. fuse
- C. shear pin or key
- D. none of the above

B 31. In order to bank a fire, the _____ is disengaged.
- A. combination fan
- B. coal feed
- C. feedwater
- D. boiler stop valve

C 32. In a screw-feed stoker, a(n) _____ draft fan supplies air for combustion.
- A. induced
- B. combination
- C. forced
- D. natural

B 33. To prevent smoke and to aid in complete combustion in the screw-feed stoker, _____ is provided using a separate damper control.
- A. underfire air
- B. overfire air
- C. retort air
- D. grate zone air

D 34. The ram-feed stoker is a(n) _____ stoker.
- A. overfeed
- B. traveling grate
- C. side feed
- D. underfeed

40 LOW PRESSURE BOILERS WORKBOOK

__D__ 35. A(n) _____ system is a solid-state control system in which a building automation controller is wired directly to control devices.
  A. ultraviolet burner control (UBC)
  B. modulating control status (MCS)
  C. programmed emission safeguard (PES)
  D. direct digital control (DDC)

__A__ 36. Combustion is the rapid burning of fuel and oxygen, resulting in _____.
  A. release of heat
  B. release of steam
  C. oxidation
  D. nitrogen and oxygen

__B__ 37. A(n) _____ system is a combustion control system that controls the amount of steam produced by starting and stopping the boiler.
  A. modulating control
  B. ON/OFF control
  C. burner proving
  D. none of the above

__B__ 38. _____ combustion is the burning of all the fuel using the minimum amount of excess air.
  A. Incomplete
  B. Complete
  C. Perfect
  D. Imperfect

__A__ 39. _____ combustion occurs when all the fuel is not burned, resulting in formation of soot and smoke.
  A. Incomplete
  B. Complete
  C. Perfect
  D. Imperfect

__D__ 40. A(n) _____ is a combustion control system that controls the amount of steam produced by changing the burner firing rate.
  A. ON/OFF control system
  B. aquastat
  C. temperature flowmeter
  D. modulating control system

__A__ 41. A _____ is burner control equipment that monitors the burner start-up sequence and the main flame during normal operation.
  A. flame safeguard system
  B. flue gas analyzer
  C. modulation control system
  D. none of the above

42. A(n) _____ is a control that functions as the mastermind of the burner control system to control the firing cycle.
_C_
   A. pressure control
   B. butterfly solenoid
   C. programmer
   D. primary combustion analyzer

43. A pollutant is matter that contaminates _____.
_D_
   A. air
   B. soil
   C. water
   D. all of the above

44. _____ is the amount of fuel the burner is capable of burning in a given unit of time.
_C_
   A. High fire
   B. Low fire
   C. Firing rate
   D. Heating surface

45. A(n) _____ automatic combustion control is most commonly used on low pressure boilers.
_A_
   A. ON/OFF
   B. positioning
   C. metering
   D. combination

46. _____ fire is burning the maximum amount of fuel in a given unit of time.
_B_
   A. Low
   B. High
   C. Maximum
   D. Medium

47. Combustion controls regulate _____.
_D_
   A. fuel supply in proportion to steam demand
   B. air supply
   C. ratio of air to the fuel supplied
   D. all of the above

48. _____ air controls the amount of fuel oil capable of being burned.
_B_
   A. Forced
   B. Primary
   C. Secondary
   D. all of the above

C____ 49. _____ air controls the combustion efficiency.
- A. Forced
- B. Primary
- C. Secondary
- D. all of the above

C____ 50. When the flame safeguard sequences the burner function, _____ is the period of time during which the pilot and main burner must be lit.
- A. prepurge
- B. pilot flame-establishing period
- C. ignition trials
- D. postpurge

B____ 51. Unburned fuel in a gaseous state is removed from the furnace by _____.
- A. vacuum pumps
- B. purging the furnace
- C. burning the fuel
- D. all of the above

B____ 52. Unburned fuel oil that is heated in the furnace will _____.
- A. turn into steam
- B. vaporize
- C. solidify
- D. flow slowly

C____ 53. A photocell sensor senses _____ light.
- A. infrared
- B. ultraviolet
- C. visible
- D. all of the above

A____ 54. The flame scanner proves the _____ and main flame.
- A. pilot
- B. combustion air
- C. high fire valve
- D. stack flame

B____ 55. Well-designed burners firing gaseous and liquid fuels operate at excess air levels of approximately _____%.
- A. .433
- B. 15
- C. 25
- D. 212

_A_ 56. The microcomputer burner control system performs all the functions of a _____.
   A. metering control
   B. positioning control
   C. ON/OFF control
   D. conventional programmer

_D_ 57. In addition to its normal functions, the MBCS also provides _____.
   A. improved combustion safety
   B. a report of all safety shutdowns
   C. increased energy conservation
   D. all of the above

_B_ 58. The primary function of the MBCS is _____.
   A. low water protection
   B. combustion safety
   C. controlling high water
   D. none of the above

_D_ 59. The MBCS can be used on automatically fired _____ single burner applications.
   A. fuel oil
   B. gas
   C. combination gas/fuel oil
   D. all of the above

_C_ 60. When an increase in steam pressure is required, the _____ activates the programmer to start the firing cycle.
   A. aquastat
   B. remote sensor
   C. pressure control
   D. modulating pressure control

Fuel Oil Pump

_B_ 1. Crescent seal
_E_ 2. Shaft
_A_ 3. Suction side
_F_ 4. Drive gear
_C_ 5. Ring gear
_D_ 6. Discharge side

Low Pressure Gas System

I __ 1. Butterfly valve
E __ 2. Gas pressure switch
L __ 3. Gas burner
A __ 4. Main gas shutoff cock
N __ 5. Pilot shutoff cock
G __ 6. Blower
K __ 7. Gas pilot

D __ 8. Manual reset valve
H __ 9. Mixing chamber
B __ 10. Gas supply line
M __ 11. Main gas solenoid valve
C __ 12. Pilot solenoid valve
J __ 13. Venturi
F __ 14. Gas pressure regulator

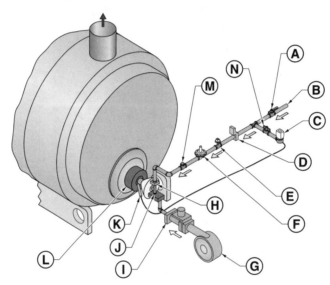

Screw-Feed Stoker

F __ 1. Feed screw
I __ 2. Apron
B __ 3. Cutoff gate
D __ 4. Pusher block
H __ 5. Motor
J __ 6. Extension
C __ 7. Retort
G __ 8. Transmission
A __ 9. Hopper
E __ 10. Dumping grates

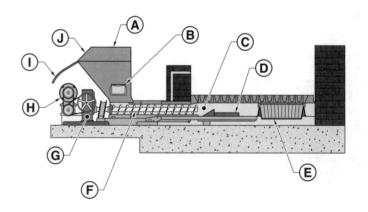

Fuel System 45

High Pressure Gas System

L	1. Utility meter	G	8. Pilot pressure regulator
E	2. Pilot vent valve	B	9. Pilot valves
I	3. Main gas vent valve	F	10. Pilot pressure gauge
A	4. Pilot adjusting cock	K	11. Main gas shutoff cock
M	5. Utility pressure regulating valve	C	12. Pilot shutoff cock
J	6. Low gas pressure switch	H	13. Main gas valve
D	7. Butterfly gas valve	N	14. Plant pressure regulating valve

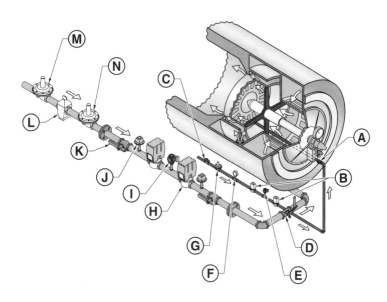

Ram-Feed Stoker

G	1. Blower
E	2. Air chamber
A	3. Retort chamber
I	4. Grate bars
D	5. Auxiliary pusher blocks
B	6. Fuel bed
H	7. Hopper
F	8. Feeder block
C	9. Sliding bottom

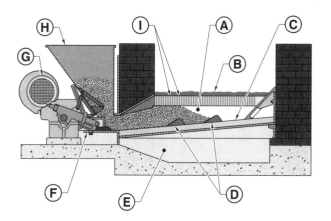

Fuel Oil Burner

- **D** 1. Rotary cup
- **A** 2. Air atomizing
- **C** 3. Pressure atomizing
- **B** 4. Steam atomizing

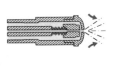 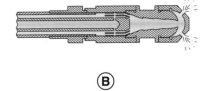 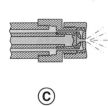 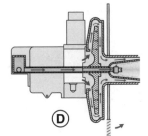

Modulating Control

- **C** 1. Bellows
- **D** 2. Pressure setting
- **B** 3. To siphon
- **A** 4. Differential setting

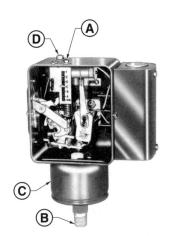

Flame Sensor

- **C** 1. Ultraviolet
- **A** 2. Flame rod
- **D** 3. Infrared
- **B** 4. Photocell

Fuel Oil System

G   1. Vacuum gauge
L   2. Fuel oil relief valve
A   3. Nozzle air pressure gauge
E   4. Fuel oil controller
J   5. Fuel oil strainer
C   6. Modulating cam
N   7. Fuel oil burner pressure gauge
H   8. Check valve
F   9. Fuel oil thermometer
K   10. Metering valve
B   11. Main fuel oil solenoid valves
M   12. Fuel oil pressure gauge
I   13. Fuel oil pump
D   14. Fuel oil pressure regulator

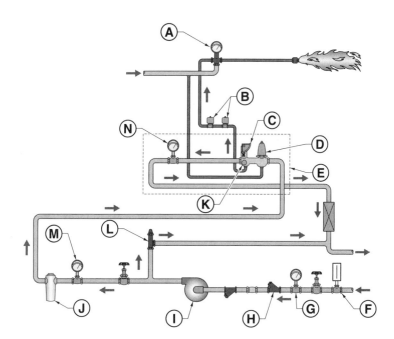

Fuel Oil Strainer

G	1. Gasket
F	2. Inlet
A	3. Strainer basket
E	4. Basket housing
I	5. Flange cover
B	6. Hand screw
H	7. Outlet
D	8. Drain plug
C	9. Plug valve
J	10. Handle position indicates basket in service

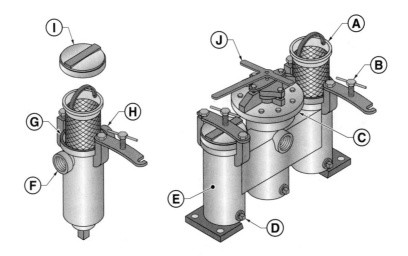

Draft System

Text Reference Pages 139 – 150

6

Name_____ Date _____

True-False

- **T** F   1. Draft is the difference in pressure between two areas that causes air or gases of combustion to flow.
- T **F**   2. Gases move from an area of low pressure to an area of high pressure. *HIGH ... LOW*
- T **F**   3. The two types of draft are mechanical and forced. *NATURAL*
- **T** F   4. Mechanical draft is produced by a fan or blower.
- **T** F   5. Air is necessary for the combustion process to take place.
- **T** F   6. Dampers are used to control the flow of air and gases of combustion.
- T **F**   7. Approximately 75 lb of air is required for every pound of fuel burned.
- **T** F   8. The amount of natural draft generated is affected by the height of the chimney.
- **T** F   9. The two types of mechanical draft are forced and induced.
- T **F** 10. The induced draft fan is located on the front of the boiler. *AFTER*
- **T** F 11. Extraordinary changes that occur in the oxygen content of the gases of combustion indicate a potential problem in the fuel system and/or draft system.
- **T** F 12. When using natural draft, the boiler is limited in the amount of fuel that can be burned.
- T **F** 13. The forced draft fan is located in the breeching. *INDUCED*
- **T** F 14. Some boilers have combination forced and induced draft fans.
- T **F** 15. Draft gauges are calibrated the same as pressure gauges. *INCHES OF WATER*
- **T** F 16. A cool stack condition can result in condensation of water vapor in the gases of combustion.
- **T** F 17. Natural draft produces greater amounts of draft in the winter than in the summer.
- **T** F 18. When using a manometer to measure draft, the liquid level in one leg is compared to the liquid level in the other leg.

BOILER ASSEMBLY

50 LOW PRESSURE BOILERS WORKBOOK

T F **19.** Proper control of draft results in higher combustion efficiency.
T F **20.** The amount of draft determines the rate of combustion.

Multiple Choice

B _____ **1.** The induced draft fan is located in the _____.
   A. boiler room
   B. breeching
   C. chimney
   D. fan room

A _____ **2.** The amount of natural draft present can be affected by _____.
   A. dampers opened or closed
   B. temperature of the gases of combustion
   C. height of the chimney
   D. all of the above

D _____ **3.** Draft is the difference in pressure between two points that causes _____ to flow.
   A. air
   B. gas
   C. gases of combustion
   D. all of the above

D _____ **4.** A _____ is a steel chimney used to direct the flow of gases of combustion from the boiler to the atmosphere.
   A. damper tube
   B. tube sheet
   C. refractory assembly
   D. stack

B _____ **5.** Forced draft and induced draft are two types of _____ draft.
   A. natural
   B. mechanical
   C. induced
   D. all of the above

B _____ **6.** Mechanical draft is produced by _____.
   A. a chimney
   B. power-driven fans
   C. temperature difference
   D. all of the above

A _____ 7. _____ draft is air that is pulled through the boiler.
- A. Induced
- B. Forced
- C. Combination
- D. Natural

B _____ 8. _____ draft is produced when air is pushed through the burner.
- A. Induced
- B. Forced
- C. Combination
- D. Natural

D _____ 9. When using _____ draft, the boiler is limited in the amount of fuel that can be burned.
- A. induced
- B. forced
- C. combination
- D. natural

A _____ 10. By using _____ draft, higher rates of combustion can be achieved.
- A. mechanical
- B. natural
- C. manufactured
- D. atmospheric

D _____ 11. Cold outside air affects the amount of _____ draft produced.
- A. induced
- B. forced
- C. combination
- D. natural

B _____ 12. Draft gauge measurements are expressed in inches of _____.
- A. mercury
- B. water column
- C. saturated steam
- D. all of the above

C _____ 13. A _____ is a simple draft gauge consisting of a U-shaped glass tube.
- A. pyrometer
- B. dynometer
- C. manometer
- D. trynometer

B   14. When measuring draft in the boiler, one leg of the tube is open to the boiler and the other to the _____.
- A. windbox
- B. atmosphere
- C. forced draft fan
- D. induced draft fan

A   15. Improperly burned fuel results in _____.
- A. soot and smoke
- B. high surface tension
- C. low chimney temperature
- D. increased steam consumption

B   16. High combustion efficiency produces the maximum amount of _____ generated by the fuel.
- A. carbon monoxide
- B. heat
- C. water
- D. ash

A   17. Too much draft when burning coal may cause _____.
- A. too hot a fire
- B. low steam pressure
- C. high water
- D. reduced fuel consumption

C   18. Combination forced and induced draft is also known as _____ draft.
- A. natural
- B. positive flow
- C. balanced
- D. forced natural

Manometer

C   1. Positive reading

A   2. No reading

B   3. Negative reading

Natural Draft

- E   1. Cold air for combustion
- C   2. Hot gases of combustion
- A   3. Gases of combustion released to atmosphere
- F   4. Cold ambient air
- B   5. Chimney
- D   6. Heat

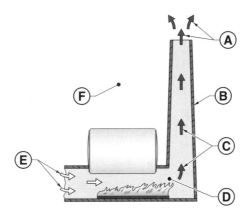

Forced Draft

- F   1. Forced draft fan
- J   2. Chimney
- A   3. Gases of combustion
- G   4. Air flow
- I   5. Boiler drum

- B   6. Outlet damper
- H   7. Furnace
- D   8. Air entering furnace
- E   9. Inlet damper
- C   10. Breeching

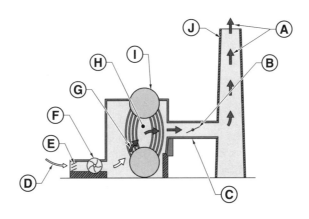

Draft Control

E _____ 1. Burner nozzle
B _____ 2. Blower motor
H _____ 3. Gases of combustion
G _____ 4. Outlet
C _____ 5. Forced draft fan
F _____ 6. Boiler
D _____ 7. Gas pilot assembly
A _____ 8. Intake

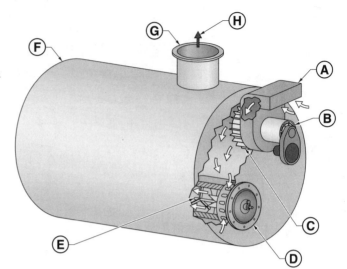

Stack

B _____ 1. Stack
G _____ 2. Boiler
D _____ 3. Drain connection
C _____ 4. Cleanout
F _____ 5. Forced draft fan
E _____ 6. Breeching
A _____ 7. Gases of combustion to atmosphere
H _____ 8. Outlet
I _____ 9. Offset stack

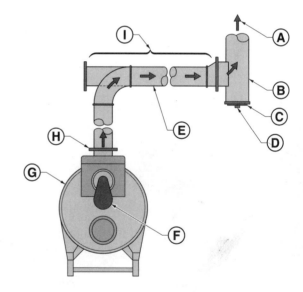

Boiler Water Treatment
Text Reference Pages 151 – 174

Name_____ Date_____

True-False

(T)	F	1. Boiler water treatment is required for makeup water introduced to the boiler.
T	(F)	2. Slug feeding is used to generate foam to prevent carryover. *feeds chemicals to boiler*
(T)	F	3. Dissolved solids are impurities such as calcium, silica, and iron dissolved in solution.
(T)	F	4. Priming and carryover can lead to water hammer and possible pipe rupture.
T	(F)	5. Oxygen in the boiler water is used to remove nonadhering sludge. *causes corrosion + pitting*
T	(F)	6. Water that contains a ~~small~~ *large* quantity of minerals is hard water.
(T)	F	7. Scale acts as an insulator and slows down the transfer of heat to the water.
(T)	F	8. Chemicals added to the boiler water change scale-forming salts into nonadhering sludge.
T	(F)	9. Sludge is removed from the boiler using the surface blowdown valve. *bottom*
(T)	F	10. A boil-out procedure is performed when a new boiler is installed.
T	(F)	11. Well water requires chemical treatment but city water does not.
T	(F)	12. The bypass feeder is located on the main header to provide efficient feeding. *discharge side of boiler feedwater pump*
(T)	F	13. Priming and carryover can be caused by opening the main steam valve too quickly.
T	(F)	14. Corrosion is the accumulation of sludge on boiler heating surfaces. *rusting of boiler metal*
(T)	F	15. Fuel oil contamination of the boiler water can be prevented by dumping all fuel oil heater returns to waste.
T	(F)	16. Scale and sediment cause pitting of the boiler metal surfaces. *oxygen*
(T)	F	17. Sodium sulfite is an oxygen scavenger that is commonly used to treat boiler water.
T	(F)	18. Automatic valves are used to control the amount of chemicals added by the bypass feeder. *hand operated*
(T)	F	19. Rusting of the boiler metal is caused by oxygen in the boiler water.

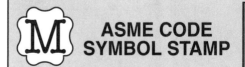

56 LOW PRESSURE BOILERS WORKBOOK

T **F** 20. Sludge rises when the boiler is on light load and is removed using the try cocks on the water column at the NOWL. *LOWERS* / *BOTTOM BLOWDOWN*

T F 21. Scale formation can result in an increase in fuel required to generate steam.

T F 22. Boiler water treatment required for most low pressure boilers is minimal if the boiler recovers all or most of the condensate returns.

T **F** 23. The chemicals used to treat the boiler water depend on the pressure inside the boiler. *DEPENDS ON WATER*

T F 24. Fuel oil contamination of the boiler water most commonly occurs in plants using No. 6 fuel oil.

T F 25. Oxygen in the boiler water can be removed by adding sodium sulfite.

Multiple Choice

B 1. Water that contains large quantities of minerals is called _____.
- A. soft water
- B. hard water
- C. sludge
- D. sediment

C 2. Small particles of water carried into steam lines are called _____.
- A. scale formation
- B. blistering
- C. carryover
- D. priming

B 3. _____ is a condition caused when steam bubbles are trapped below the boiler water surface.
- A. Sludge slugging
- B. Foaming
- C. Priming
- D. Tube corrosion

B 4. _____ is the accumulation of high alkaline elements, which cause boiler metal corrosion at stress zones.
- A. Sodium sulfate
- B. Caustic embrittlement
- C. Corrosion
- D. Pitting

A 5. _____ is used as an oxygen scavenger to remove oxygen from the boiler water.
- A. Sodium sulfite
- B. Zeolite
- C. Sodium sulfate
- D. none of the above

A

6. Nonadhering sludge is best removed when the boiler is _____.
   A. under light load
   B. under heavy load
   C. being drained
   D. being tested

D

7. Overheating of heating surfaces results in _____.
   A. bags
   B. blisters
   C. burned out tubes
   D. all of the above

D

8. In low pressure boiler operation, chemicals for reducing oxygen and preventing scale can be added using a _____.
   A. vacuum pump
   B. blowdown valve
   C. safety valve
   D. bypass feeder

D

9. Scale is caused by _____.
   A. scale-forming salts
   B. hard water
   C. mineral deposits
   D. all of the above

D

10. _____ in the boiler water causes corrosion and pitting of the boiler metal.
    A. Sodium sulfite
    B. Sodium sulfate
    C. Carbon dioxide
    D. Oxygen

D

11. High surface tension on the boiler water is shown by _____.
    A. pressure above 15 psi
    B. steam released by the top try cock
    C. increased amounts of sludge
    D. fluctuation in the gauge glass

A

12. Water is _____ to reduce the amount of oxygen.
    A. heated
    B. cooled
    C. pressurized to 15 psi
    D. all of the above

A

13. Increased surface tension of boiler water can be caused by _____.
    A. fuel oil inside the boiler
    B. superheated steam
    C. excessive pressure in the safety valve
    D. water in the condensate

D _____ 14. Fuel oil is removed from the steam and water side of a boiler by using _____.
   A. sulfuric acid
   B. sulfate
   C. silica gel
   D. caustic soda

D _____ 15. _____ can lead to water hammer.
   A. Priming
   B. Carryover
   C. High surface tension
   D. all of the above

C _____ 16. Nonadhering sludge is removed from the boiler by using the _____.
   A. bottom try cock
   B. gauge glass blowdown valve
   C. bottom blowdown valve
   D. surface blowdown valve

A _____ 17. Hard water contains _____.
   A. scale-forming salts
   B. makeup chemicals
   C. oxygen scavengers
   D. sodium sulfite

Sludge Removal

D _____ 1. To blowdown line
A _____ 2. NOWL
C _____ 3. Bottom blowdown valve
E _____ 4. Boiler water circulation reduced
B _____ 5. Sludge

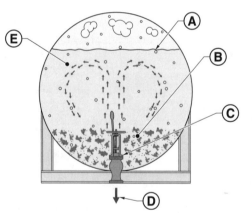

Priming and Carryover

1. D — Boiling water
2. A — Water carried into steam lines
3. B — Feedwater
4. E — High water level
5. C — Boiler

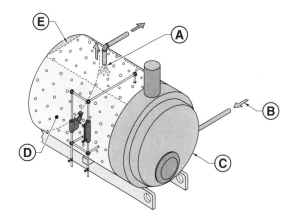

Foaming

1. B — Trapped steam bubbles
2. E — Boiler
3. D — Boiler water
4. A — Film from impurities
5. C — Steam bubbles

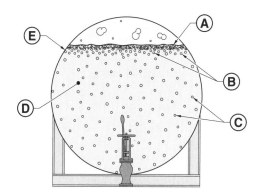

Water Treatment Log Readings

1. 8.9 — Third day of month – condensate pH
2. 45 — Sixth day of month – sodium sulfite present
3. 16 — Second day of month – condensate TDS
4. 0 — Fifth day of month – condensate hardness
5. 2700 — Fourth day of month – boiler water TDS

Bypass Feeder

E 1. Bypass feeder tank
B 2. Feedwater pump
D 3. Stop valve
A 4. Water treatment chemicals added
C 5. Check valve
F 6. Boiler

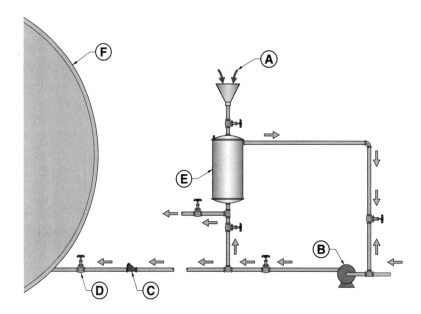

Boiler Operation Procedures

Text Reference Pages 175 – 202

Name_____ Date_____

True-False

- (T) F 1. When taking over a shift, preliminary safety checks allow the boiler operator to identify any possible problems.
- (T) F 2. A routine set of duties helps the operator to complete all assigned tasks.
- (T) F 3. In general, boilers should be blown down every 24 hr.
- (T) F 4. The main steam stop valve should be closed during normal plant start-up.
- T (F) 5. The boiler vent must be closed during warm-up to prevent pressure from escaping. *OPEN*
- (T) F 6. A bottom blowdown can be used to control chemical concentrations in the boiler.
- T (F) 7. The furnace brickwork should be cooled rapidly for energy efficiency. *SLOWLY TO PREVENT SPALLING (FLAKING)*
- T (F) 8. There should be pressure in the boiler when dumping. *NO*
- T (F) 9. The boiler should be blown down when it is on a heavy load. *LIGHT*
- (T) F 10. The operator's hand must always be kept on the blowdown valve during blowdown.
- T (F) 11. The blowdown valve should be kept open long enough for the water level in the gauge glass to drop out of sight. *NEVER*
- (T) F 12. The quick-opening valve is opened first and closed last during blowdown.
- (T) F 13. The low water fuel cutoff is checked for proper operation with the burner firing.
- (T) F 14. Try cocks can be used as a boiler water level indicator.
- T (F) 15. Steam discharged from the top try cock indicates a low water level. *STARVED*
- T (F) 16. An evaporation test is performed by blowing down the low water fuel cutoff. *STOP FEEDWATER / LET CONDENSATE BOILER*
- (T) F 17. Water added to a boiler with a low water level condition could cause a boiler explosion.
- T (F) 18. Furnace explosions are caused by steam in the fuel oil. *COMBUSTIBLE GASES OR VAPORS IN FURNACE*
- (T) F 19. Any abnormal odor of gas should alert the boiler operator to a possible leak.

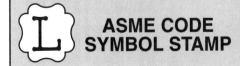

ASME CODE SYMBOL STAMP — LOCOMOTIVE BOILER

20. The boiler must be filled with water heated to 50°F before it is started for the first time. [F] *(low fire / slowly 70°F?)*
21. Bottom blowdown is performed by opening the gauge glass blowdown valve. [F] *(bottom of boiler)*
22. Leaking fuel oil in the furnace could lead to a furnace explosion. [T]
23. The boiler should be inspected to determine the cause of a low water level condition. [T]
24. The furnace should not be purged after a flame failure. [F]
25. A boiler must be taken off-line before it can be inspected. [T]
26. Main steam stop valves must be tagged out during a boiler inspection. [T]
27. After a boiler is dumped, it should dry completely before cleaning. [F] *(wash out ASAP / easier)*
28. A hydrostatic test is a steam test used to check for boiler leaks. [F] *(water pressure test)*
29. A malfunctioning steam trap can result in a steambound feedwater pump. [T]
30. Tools and equipment must be provided for the boiler inspector during boiler inspection. [T]
31. During a hydrostatic test, the boiler is half-filled with water. [F] *(completely filled)*
32. Dry lay-up is recommended for boilers that will be out of service for an extended period. [T]
33. Moisture left on metal surfaces during dry lay-up could cause scale buildup. [F] *(corrosion)*
34. More pounds of silica gel than quicklime are required when laying up a boiler dry. [T]
35. All valves are opened during dry lay-up. [F] *(closed)*
36. A scratch on the inside of a gauge glass can cause it to break. [T]
37. A clean gauge glass is necessary to accurately determine boiler water level. [T]
38. A boiler inspection is commonly performed by the plant manager. [F] *(local boiler inspector)*
39. An evaporation test lowers the boiler water level rapidly. [F] *(gradually + evenly)*
40. Two people should be present during an evaporation test. [T]
41. All plants maintain boiler room logs for an 8-hour period. [F] *(some)*
42. A boiler room log can be used to determine the cause of a boiler shutdown. [T]
43. Boiler lay-up is required when the boiler is out of service between shifts. [F] *(for extended length of time)*
44. When warming up the boiler, the low water fuel cutoff should be blown down before the steam pressure gauge records pressure. [F] *(when burner is firing)*
45. When shutting down the boiler, the main steam valve should be closed immediately after the furnace is shut down. [F] *(after boiler has stopped producing steam)*

Multiple Choice

_D_ 1. The first thing a boiler operator should do when taking over a shift is _____.
   A. read the boiler room log
   B. check the fuel supply
   C. blow down the flash tank
   D. check the boiler water level

_C_ 2. When the low water fuel cutoff is blown down, _____.
   A. the water in the gauge glass will rise
   B. foaming is increased
   C. the burner should shut off
   D. none of the above

_B_ 3. The boiler should always be _____ the header pressure when cutting a boiler in on the line.
   A. slightly above
   B. slightly below
   C. the same as
   D. none of the above

_C_ 4. The boiler should be blown down when it is _____.
   A. in high fire
   B. shut down
   C. on a light load
   D. all of the above

_B_ 5. ASME code states that on boilers having two bottom blowdown valves, the _____ valve should be closest to the boiler.
   A. screw
   B. quick-opening
   C. os&y
   D. globe

_B_ 6. The gauge glass on the boiler must be located so the lowest visible part of the gauge glass is _____ above the highest heating surface.
   A. 1″ to 2″
   B. 2″ to 3″
   C. 3″ to 4″
   D. 4″ to 5″

_D_ 7. If water cannot be seen in the gauge glass, _____.
   A. add water immediately
   B. the boiler inspector should be notified
   C. the fusible plug must be inspected
   D. the burner should be secured

64 LOW PRESSURE BOILERS WORKBOOK

D _____ 8. A furnace explosion can be caused by _____.
- A. a buildup of combustible gases or vapors
- B. leaking gas or fuel oil
- C. improper furnace purge
- D. all of the above

A _____ 9. The _____ blowdown valve is used to dump the boiler.
- A. bottom
- B. surface
- C. water column
- D. all of the above

C _____ 10. The best method for checking for leaks in gas lines is _____.
- A. smelling the area
- B. lighting a match
- C. applying soapy water
- D. none of the above

D _____ 11. A(n) _____ test is used to check for leaks in the boiler.
- A. evaporation
- B. low water
- C. caustic
- D. hydrostatic

C _____ 12. All traces of coal, soot, and ash must be removed to prevent _____ from forming when mixed with water.
- A. sodium sulfite
- B. zeolite
- C. sulfuric acid
- D. silica gel

A _____ 13. A steambound pump is caused by _____.
- A. excessive water temperature
- B. improper feedwater treatment
- C. low furnace temperature
- D. none of the above

B _____ 14. A(n) _____ test is used to test the low water fuel cutoff.
- A. hydrostatic
- B. evaporation
- C. dry lay-up
- D. vacuum

15. A _____ is used to record information regarding operation of the boiler during a given period of time. _D_
    A. chief instruction handbook
    B. pressure control
    C. pressure gauge
    D. boiler room log

16. The boiler must be _____ when the low water fuel cutoff is blown down. _C_
    A. shut down
    B. dumped
    C. firing
    D. in battery

17. When the flame scanner is removed with the burner firing, the _____. _C_
    A. fuel valve should open
    B. programmer should start the firing cycle
    C. burner should shut off
    D. all of the above

18. To prevent spalling of the furnace brickwork, the furnace _____. _A_
    A. should be cooled slowly
    B. is lit off in high fire
    C. is purged before shutdown
    D. none of the above

19. Boiler _____ is a condition when a boiler is operating at or above its maximum allowable working pressure. _C_
    A. NOWL
    B. MAWP
    C. overpressure
    D. superheating

20. Before dumping the boiler, the boiler heating surface should be _____. _B_
    A. in high fire
    B. cool enough to touch
    C. higher than 337°F
    D. in low fire

21. The water column whistle valve is _____ during the hydrostatic test. _A_
    A. removed
    B. activated once
    C. sounded
    D. none of the above

A 22. During the hydrostatic test, pressure on the boiler is brought up to _____ times the MAWP.
  A. 1½
  B. 2
  C. 2½
  D. 5

D 23. Water used to fill the boiler during the hydrostatic test should be a minimum of _____ °F.
  A. 0
  B. 12
  C. 32
  D. 70

B 24. _____ is placed on the water side of the boiler when placing the boiler in dry lay-up.
  A. Anthracite
  B. Quicklime
  C. Sulfite
  D. none of the above

D 25. Gauge glass nuts are tightened by hand and then turned ¼ TURN with a wrench.
  A. ¾ of a turn
  B. 1½ turns
  C. 2 turns
  D. none of the above

D 26. A EVAPORATION test is the most accurate method of testing the low water fuel cutoff.
  A. surface blowdown
  B. hydrostatic
  C. vacuum
  D. none of the above

A 27. When testing a low water fuel cutoff, the burner should shut off when _____.
  A. water is still visible in the gauge glass
  B. the gauge glass is empty
  C. the float is removed
  D. all of the above

A 28. When performing a bottom blowdown, the valves should be opened and closed _____.
  A. slowly
  B. quickly
  C. at the same time
  D. when water is not present in the gauge glass

C_____ 29. To correct a high water condition, a _____ is performed.
   A. low water fuel cutoff test
   B. surface blowdown
   C. bottom blowdown
   D. boiler vent test

D_____ 30. New _____ should be used when replacing a broken gauge glass.
   A. packing nuts
   B. glands
   C. seats
   D. washers

D_____ 31. A boiler is given a bottom blowdown to _____.
   A. discharge sludge
   B. control high water
   C. control chemical concentrations
   D. all of the above

A_____ 32. A leak on gas lines and equipment must be repaired _____.
   A. by authorized personnel
   B. after the shift by the boiler operator
   C. only if the load is interrupted
   D. while the burner is still firing

A_____ 33. For maximum safety in operation of a low pressure boiler, the low water fuel cutoff should be tested _____.
   A. daily
   B. weekly
   C. monthly
   D. annually

D_____ 34. Boiler data that is commonly recorded on a boiler room log includes _____.
   A. boilers on line
   B. steam pressure
   C. condensate return temperature
   D. all of the above

D_____ 35. All running auxiliaries (fuel oil pumps, fan, water pump, and burner) should be checked for proper _____.
   A. temperature
   B. pressure
   C. lubrication
   D. all of the above

C _____ 36. The proper testing of boiler accessories is determined by procedures suggested by _____.
- A. MAWP
- B. ABMA regulations
- C. ASME code
- D. all of the above

B _____ 37. After testing the flame scanner, _____.
- A. the burner should go to high fire
- B. reset the programmer
- C. blow down the fuel oil return
- D. remove the burner tip

B _____ 38. During the evaporation test, the _____ is secured.
- A. fuel to the burner
- B. automatic city water makeup feeder
- C. gauge glass inlet
- D. all of the above

B _____ 39. During a boiler inspection, the main steam stop valve is _____.
- A. opened and tagged out
- B. closed and tagged out
- C. opened partially to allow steam flow
- D. none of the above

A _____ 40. The _____ must be covered with water before water can be added safely.
- A. heating surface
- B. top try cock
- C. fuel oil valve
- D. vaporstat

Hot Water Heating Systems

Text Reference Pages 203 – 224

9

Name _____ Date _____

True-False

(T) F **1.** Hot water heating systems produce heat more consistently than steam heating systems.

T **(F)** **2.** Hot water heating systems require less water than steam heating systems. *more (50 times more)*

T **(F)** **3.** The natural circulation hot water heating system uses circulating pumps to transport water in the system. *forced*

(T) F **4.** Water becomes lighter as it is heated.

T **(F)** **5.** The natural circulation hot water heating system uses a compression tank to absorb pressure changes. *forced*

T **(F)** **6.** Natural circulation hot water heating systems are commonly used in large contemporary buildings. *forced*

T **(F)** **7.** Floor-mounted circulating pumps are used in small plants. *large*

(T) F **8.** Line-mounted circulating pumps operate intermittently, depending on plant requirements.

T **(F)** **9.** The forced circulation hot water heating system is vented to the atmosphere. *natural (forced not vented)*

(T) F **10.** The aquastat controls the starting and stopping of the burner on a hot water boiler.

(T) F **11.** The flow control valve prevents natural circulation when water is not pumped in the system.

T **(F)** **12.** The pressure-reducing valve is used to prevent overpressure of water supplied to the expansion tank. *reduces incoming city water pressure / provide water to heating unit*

T **(F)** **13.** Diverter fittings are used to vent air from the heating units.

(T) F **14.** Stop valves before and after the pressure-reducing valve allow servicing of the pressure-reducing valve.

(T) F **15.** The three-way mixing valve blends water returning from the heating units with supply water from the boiler.

ASME CODE SYMBOL STAMP   WATER HEATER

(T) F 16. All boilers are manufactured in conformance with Section I or IV of the ASME Boiler and Pressure Vessel Code.

T **(F)** 17. An aquastat controls the burner by sensing steam pressure in a hot water boiler.   *TEMP. OF WATER*

(T) F 18. The strainer in the pressure-reducing valve should be inspected periodically for scale formation.

T **(F)** 19. The compression tank is normally full of water to feed into the boiler during a low water condition.   *1/2*

(T) F 20. The safety relief valve should be located on the highest part of the boiler.

(T) F 21. Safety relief valves should be manually tested every 30 days as recommended by the ASME Code.

T **(F)** 22. Safety relief valves should discharge into the ~~return line feeding into the circulating pump~~.   *AN OPEN DRAIN*

T **(F)** 23. A temperature-pressure gauge indicates temperature in the compression tank.   *OF WATER LEAVING BOILER*

(T) F 24. Hydrostatic pressure is expressed in pounds per square inch (psi).

T **(F)** 25. Air controls are used to allow water to drain from the compression tank.   *REMOVE AIR FROM SYSTEM*

(T) F 26. Air removed from the boiler water is diverted to the compression tank in a forced circulation hot water heating system.

(T) F 27. An aquastat commonly has a temperature differential preset by the manufacturer.

T **(F)** 28. The aquastat temperature differential setting controls the starting and stopping of the ~~circulating pump~~.   *BURNER*

(T) F 29. Two diverter fittings may be required for heating units above the supply line where resistance to circulation is high.

(T) F 30. Hot water heating systems require special hot water boiler accessories.

Multiple Choice

A _____ 1. Water conveys approximately _____ the amount of heat that steam conveys.
   A. one-fifth
   B. one-fourth
   C. one-third
   D. one-half

B _____ 2. A hot water heating system operates at _____ the steam heating system.
   A. a higher temperature than
   B. a lower temperature than
   C. the same temperature as
   D. all of the above

_A_ 3. In a natural circulation hot water heating system, the _____ functions as a relief valve.
   A. supply line
   B. return line
   C. heating unit
   D. expansion tank

_C_ 4. _____ pressure is water pressure per vertical foot exerted at the base of a column of water.
   A. Normal operating water
   B. Electrostatic
   C. Hydrostatic
   D. none of the above

_C_ 5. In a remote temperature-monitoring system, the burner is controlled based on _____.
   A. steam pressure
   B. temperature of air at the heating unit
   C. outside temperature
   D. makeup water

_B_ 6. Steam boilers are classified as high pressure if steam pressure exceeds _____ psi.
   A. 7
   B. 15
   C. 20
   D. 160

_C_ 7. A flow control valve functions similar to a(n) _____ valve to prevent natural water circulation.
   A. air separator
   B. pressure reducing
   C. check
   D. gate

_D_ 8. Hot water boilers operating with a 250°F water temperature and _____ psi water pressure or less are classified as low pressure.
   A. 10
   B. 15
   C. 100
   D. 160

_D_ 9. Hot water boilers can be _____ boilers.
   A. firetube
   B. watertube
   C. cast iron sectional
   D. all of the above

72 LOW PRESSURE BOILERS WORKBOOK

D 10. A hot water boiler safety relief valve is rated in _____ relieved per hour.
   A. Btu of steam temperature
   B. pounds of water pressure
   C. pounds of water temperature
   D. none of the above

B 11. A steam boiler safety valve is rated in _____ relieved per hour.
   A. Btu of steam temperature
   B. pounds of steam
   C. Btu of water temperature
   D. all of the above

A 12. A safety relief valve must be built according to the _____.
   A. ASME Code
   B. ABMA Code
   C. boiler manufacturer specifications
   D. none of the above

C 13. The relieving pressure of a safety relief valve on a low pressure hot water boiler cannot exceed _____ psi.
   A. 15
   B. 30
   C. 160
   D. 250

C 14. Hot water boilers usually have _____ psi on the boiler at all times.
   A. 3 to 5
   B. 10 to 15
   C. 12 to 18
   D. 160 to 250

B 15. The _____ tank is normally half full of water to maintain the correct water level in the forced circulation hot water heating system.
   A. expansion
   B. compression
   C. receiver
   D. vacuum

C 16. Section VI of the ASME Code recommends safety relief valves be manually tested _____.
   A. once a shift
   B. every 24 hours
   C. every 30 days
   D. none of the above

17. The _____ protects the hot water heating system from exceeding the MAWP. **[C]**
    A. aquastat
    B. blowdown valve
    C. safety relief valve
    D. compression tank

18. The _____ controls the starting and stopping of the burner by sensing the temperature in the hot water boiler. **[D]**
    A. flow control valve
    B. diverter fitting
    C. circulating pump
    D. aquastat

19. The temperature-pressure gauge indicates the temperature and pressure of the water _____. **[B]**
    A. in the heating unit
    B. leaving the boiler
    C. in the compression tank
    D. at the mixing valve

20. _____ direct the flow of hot water supplied to and returned from the heating units. **[D]**
    A. Flow control valves
    B. Hand-operated valves
    C. Steam traps
    D. Diverter fittings

21. A _____ prevents natural circulation when hot water is not being pumped through the system. **[D]**
    A. stop valve
    B. diverter fitting
    C. circulating pump
    D. flow control valve

22. A pressure-reducing valve reduces the pressure of city water to approximately _____ psi. **[D]**
    A. 4 to 6
    B. 8 to 10
    C. 10 to 15
    D. 12 to 18

23. A(n) _____ allows bleeding of trapped air in a heating unit. **[C]**
    A. steam trap
    B. diverter fitting
    C. air vent
    D. air control

74 LOW PRESSURE BOILERS WORKBOOK

__B__ 24. If the compression tank is allowed to fill with water, pressure will increase, causing the _____.
   A. pressure-reducing valve to activate
   B. safety relief valve to open
   C. air control to drain the water
   D. all of the above

__C__ 25. Larger compression tanks are equipped with a _____ for determining the compression tank water level.
   A. sight glass
   B. try cock
   C. gauge glass
   D. none of the above

Natural Circulation Hot Water Heating System

__G__ 1. Vent line
__A__ 2. Heating unit
__H__ 3. Overflow line
__C__ 4. Boiler
__F__ 5. Expansion tank
__E__ 6. Branch line
__B__ 7. Return line
__D__ 8. Supply line

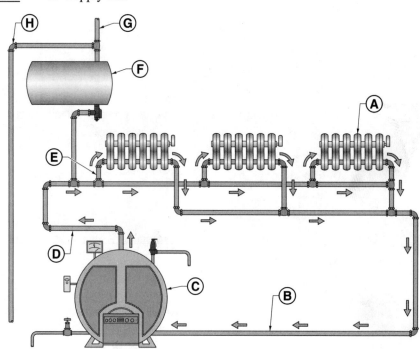

Forced Circulation Hot Water Heating System

K 1. Air separators
A 2. Heating unit
Q 3. Temperature-pressure gauge
H 4. Drain
T 5. Compression tank
E 6. Backflow preventer
O 7. Boiler
S 8. Compression tank valve
B 9. Air vent
P 10. Aquastat
F 11. Bypass valve

L 12. Stop valves
D 13. Air control tank fitting
I 14. Flow control valve
M 15. Makeup water supply line
G 16. Pressure-reducing valve
C 17. Diverter fitting
N 18. Safety relief valve
J 19. Bypass line
R 20. Circulating pump

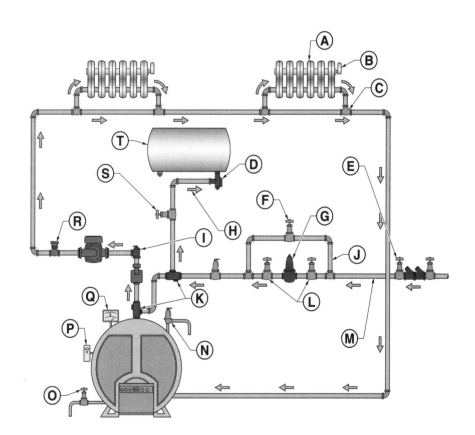

Safety Relief Valve

D 1. Sealing diaphragm
F 2. Body
B 3. Outlet
E 4. Spring
A 5. Data plate
G 6. Try lever
C 7. Inlet

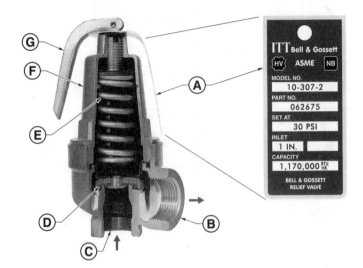

Aquastat

B 1. Air separator
G 2. Set point adjustor
F 3. Thermostat
D 4. Boiler
F 5. Cover
H 6. Remote bulb sensor
C 7. Aquastat
E 8. Burner controls
A 9. Heating units

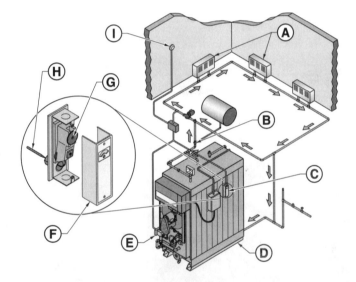

Cooling Systems

Text Reference Pages 225 – 242

10

Name _____   Date _____

True-False

- **T** F   1. Water is the most common medium used to transport heat from the area to be cooled.
- **T** F   2. Some cooling systems can be designed to share components with a hot water heating system.
- **T** F   3. Chiller systems are commonly used in refinery, brewery, and dairy applications.
- T **F**   4. Heat flows from a cold material to a hot material. *HOT TO COLD*
- **T** F   5. Heat energy can be changed to mechanical energy.
- T **F**   6. Latent heat is the amount of heat that changes the temperature of a substance but not the state of a substance. *TEMP. SAME STATE CHANGES*
- T **F**   7. When refrigerant gas is compressed, the temperature and pressure are lowered. *RAISED*
- **T** F   8. Refrigerant is converted from a gas to a liquid in the condenser.
- T **F**   9. An absorption refrigeration system uses a compressor to provide pressurized ammonia to the system. *ABSORBER/GENERATOR*
- **T** F   10. Refrigerant is changed from a liquid to a gas in the low pressure side of a compression system.
- **T** F   11. Latent heat is heat added to a substance that changes its state without a change in temperature.
- T **F**   12. In a compression refrigeration system, heat transfer occurs at the expansion valve and the compressor. *Condenser + evaporator*
- **T** F   13. When ice changes into water, there is a change in state.
- **T** F   14. In a refrigeration system, heat is absorbed when a fluid changes from a liquid to a gas.
- T **F**   15. A ton of cooling is the amount of heat required to freeze a ton of ice within a 24-hour period. *MELT*
- T **F**   16. The evaporator is located on the high pressure side of a cooling system. *LOW*

ASME CODE SYMBOL STAMP   ELECTRIC BOILER

77

Answer		Question
T	F	17. A refrigerant is a substance that absorbs heat when changing from a liquid to a gaseous state.
T	**F**	18. Direct cooling systems are used where the space or product to be cooled is located a considerable distance from the condensing equipment. *(is in direct contact)*
T	F	19. Refrigerants used in a compression refrigeration system must have a boiling point below the temperature of the air leaving the evaporator.
T	**F**	20. The most common refrigerant used in an absorption refrigeration system is lithium bromide. *(ammonia + water)*
T	F	21. In a direct cooling system, the evaporator is in direct contact with the space or product being cooled.
T	F	22. Compression refrigeration systems are commonly used for air conditioning applications.
T	**F**	23. Heat is released by the refrigerant at the evaporator. *(condenser)*
T	F	24. Liquid refrigerant leaves the condenser under high pressure before reaching the expansion valve.
T	**F**	25. Refrigerant pressure is controlled by the expansion valve located between the evaporator and the compressor. *(condenser)*
T	F	26. In a lithium bromide and water system, heat is applied at the generator.
T	**F**	27. Indirect cooling systems use refrigerant piped to the space or product to be cooled. *(chilled water)*
T	**F**	28. Chlorofluorocarbons (CFCs) are present in all refrigerants used in compression systems.
T	F	29. Cooling systems use equipment similar to hot water heating systems.
T	F	30. Air conditioning systems are cooling systems used to cool air for comfort in building spaces.

Multiple Choice

B _____ 1. Water chilled in a cooling system is chilled by the _____.
- A. boiler
- B. chiller
- C. condenser
- D. compressor

C _____ 2. Heat added to a substance that changes its state without a change in temperature is _____ heat.
- A. super
- B. sensible
- C. latent
- D. mechanical

D_____ 3. In an indirect cooling system, _____ is commonly used to absorb heat from the product or space to be cooled.
   A. ammonia
   B. lithium bromide
   C. freon
   D. water

C_____ 4. In a compression refrigeration system, liquid refrigerant under high pressure is allowed to drop in pressure by the _____.
   A. absorber
   B. absorbent
   C. expansion valve
   D. all of the above

D_____ 5. Absorption refrigeration systems may use _____ as a refrigerant.
   A. lithium bromide and water
   B. ammonia
   C. lithium chloride and water
   D. all of the above

D_____ 6. Cooling units that cool building spaces are designed to _____ heat from the air.
   A. sense
   B. release
   C. condense
   D. absorb

C_____ 7. In a refrigeration system, heat is _____ when a fluid changes from a gas to a liquid.
   A. decreased
   B. absorbed
   C. released
   D. compressed

D_____ 8. In a compression refrigeration system, high pressure vapor is converted from a gas to a liquid in the _____.
   A. evaporator
   B. compressor
   C. generator
   D. condenser

A_____ 9. In a compression refrigeration system, the refrigerant absorbs heat in the _____.
   A. evaporator
   B. compressor
   C. generator
   D. none of the above

C _____ 10. In a lithium bromide and water refrigeration system, heat is commonly applied from a source such as a steam coil at the _____.
   A. evaporator
   B. compressor
   C. generator
   D. condenser

B _____ 11. In a lithium bromide system, cooling water is directed through the _____ and condenser to remove heat.
   A. evaporator
   B. absorber
   C. generator
   D. all of the above

B _____ 12. Indirect cooling systems commonly use _____ as a medium to cool the space or product.
   A. ammonia
   B. chilled water
   C. lithium bromide
   D. Freon

A _____ 13. In a compression refrigeration system, heat is absorbed by the refrigerant in the _____.
   A. low pressure side
   B. high pressure side
   C. condenser coils
   D. generator

D _____ 14. Cooling systems are rated in _____.
   A. Btu per hour
   B. temperature changes per hour
   C. refrigerant freezing point
   D. tons of cooling

A _____ 15. In a compression refrigeration system, the _____ is used to change the pressure of a refrigerant gas.
   A. compressor
   B. absorber
   C. condenser
   D. steam coil

B _____ 16. The compressor will be damaged if _____ refrigerant is allowed to enter.
   A. gaseous
   B. liquid
   C. CFC
   D. all of the above

__D__ 17. Refrigerants used in a compression system must be _____.
   A. noncorrosive
   B. chemically stable
   C. nonflammable
   D. all of the above

__B__ 18. An absorption refrigeration system includes all of the following except the _____.
   A. condenser
   B. compressor
   C. evaporator
   D. generator

__C__ 19. _____ refrigerants do not contain chlorine and are considered safer for the environment.
   A. HCFC
   B. CFC
   C. HFC
   D. All of the above

__C__ 20. In indirect cooling systems, _____ used as a medium must be kept above 32°F.
   A. ammonia
   B. lithium
   C. water
   D. freon

Compression Refrigeration System

__D__ 1. Cool air
__B__ 2. Hot air
__F__ 3. Expansion valve
__E__ 4. Evaporator
__C__ 5. Compressor
__A__ 6. Condenser

Liquid Receiver

E _____ 1. Metering device
B _____ 2. Liquid receiver
A _____ 3. Evaporator
D _____ 4. Compressor
C _____ 5. Condenser

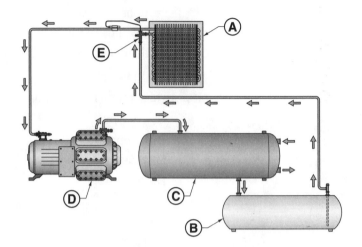

Ammonia-Water Absorption Refrigeration System

G _____ 1. Generator
B _____ 2. Expansion valve
C _____ 3. Evaporator
F _____ 4. Regulating valve
D _____ 5. Absorber
A _____ 6. Condenser
E _____ 7. Pump

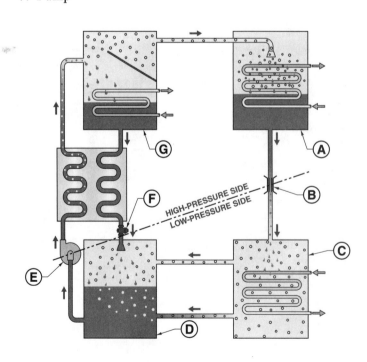

Direct Cooling System

A _____ 1. Vapor to compressor
D _____ 2. Evaporator coil
B _____ 3. Metering device
E _____ 4. Liquid from receiver
C _____ 5. Evaporator

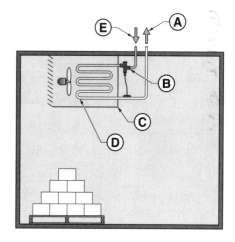

Cooling System Application – Pasteurization Process

D _____ 1. Cooled milk
B _____ 2. Warm water out
C _____ 3. Collection trough
G _____ 4. Heated milk enters cooler
F _____ 5. Milk passing around piping
A _____ 6. Chilled water in
E _____ 7. Piping containing chilled water

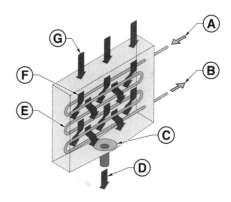

Indirect Cooling System

B _____ 1. Compressor
D _____ 2. Metering device
F _____ 3. Cooling tower
A _____ 4. Air duct
E _____ 5. Evaporator
G _____ 6. Ceiling
C _____ 7. Condenser

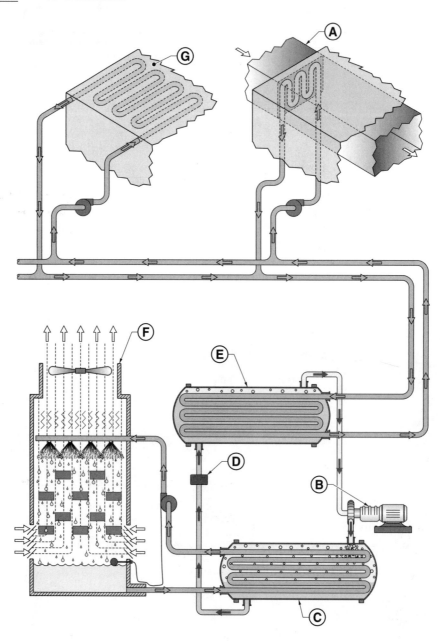

Boiler Operation Safety

Text Reference Pages 243 – 266

Name_____ Date_____

True-False

- (T) F 1. An accident can occur at any time.
- (T) F 2. All accidents should be reported regardless of the nature of the injury.
- T (F) 3. Accident report forms are filed in the boiler room log. *SEPEATE FORM*
- (T) F 4. Combustible materials require special safety precautions in handling.
- T (F) 5. Local fire departments are responsible for maintaining a safe boiler room. *INSPECTING BUILDINGS → BOILER OPERATOR*
- (T) F 6. OSHA standards are reproduced in the Code of Federal Regulations (CFR).
- T (F) 7. The NFPA is a federal agency established *FOR FIRE PROTECTION STANDARDS* ~~to control and abate pollution.~~
- (T) F 8. Fire extinguishers are identified by the type of fire that the extinguisher is designed to be used on.
- (T) F 9. Eye protection is required when visually inspecting the furnace fire.
- (T) F 10. An insurance agency may dictate special trim and accessories for maximum safety in a particular plant.
- (T) F 11. All unsafe conditions in the boiler room should be reported to the immediate supervisor.
- T (F) 12. The ABMA is a government regulatory agency. *NFPA*
- (T) F 13. When testing safety valves, stand clear to avoid possible injury.
- T (F) 14. Horseplay in the boiler room is permissible away from the boiler furnace. *NEVER*
- T (F) 15. To save time, always run in the event of an emergency. *MOVE QUICKLY + WITH PURPOSE   NEVER*

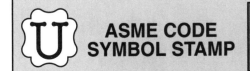

86 LOW PRESSURE BOILERS WORKBOOK

T F 16. Spontaneous combustion can be caused by improper storage of oily rags.

T F 17. All facilities must have a fire safety plan.

T F 18. A permit-required confined space requires a signed entry permit by the entry supervisor.

Multiple Choice

__D__ 1. Accident reports include _____.
- A. name of injured person
- B. date, time, and place of accident
- C. nature of duty
- D. all of the above

__B__ 2. A(n) _____ is printed material used to relay chemical hazard information from the manufacturer to the employer.
- A. OSHA CFR
- B. MSDS
- C. NFPA bulletin
- D. EPA certificate

__D__ 3. The _____ is responsible for the safe and efficient operation of the boiler.
- A. chief engineer
- B. fire brigade
- C. head custodian
- D. boiler operator

__A__ 4. When the boiler is removed from service, the first thing to do is _____.
- A. tag the steam stop valves
- B. coat the tubes with oil to prevent rusting
- C. drill drain holes to ensure removal of water
- D. all of the above

__C__ 5. The _____ uses a four-color diamond-shaped sign to display basic information about hazardous material.
- A. RTK label
- B. HMIG label
- C. NFPA Hazard Signal System
- D. none of the above

__B__ 6. _____ are used to store combustible liquids.
- A. Flash tanks
- B. Approved safety cans
- C. Vacuum pumps
- D. all of the above

7. The _____ should be opened to prevent a vacuum in the boiler before opening a manhole. **[C]**
   A. feedwater pump
   B. vacuum tank
   C. boiler vent
   D. blowdown

8. _____ is required to start and sustain a fire. **[D]**
   A. Heat
   B. Fuel
   C. Oxygen
   D. all of the above

9. Boiler valves should be opened _____ to prevent water hammer. **[B]**
   A. quickly
   B. slowly
   C. rapidly and often
   D. none of the above

10. Before starting up any equipment _____. **[D]**
    A. ask the previous operator about any special problems
    B. never take any function for granted
    C. personally check for proper function
    D. all of the above

11. Ear protection devices are rated for noise reduction with a(n) _____ number. **[A]**
    A. NRR
    B. EPA
    C. OSHA
    D. ASME

12. A(n) _____ is required for protection against airborne contaminants when cleaning boiler refractory. **[C]**
    A. hard hat
    B. ear protection device
    C. respirator
    D. none of the above

13. The number and type of fire extinguishers needed are determined by _____. **[D]**
    A. how fast the fire may spread
    B. potential heat intensity
    C. accessibility to fire
    D. all of the above

88 LOW PRESSURE BOILERS WORKBOOK

_A_   14. When blowing down the boiler, the quick-opening valve should be _____.
        A. opened first and closed last
        B. opened after the screw valve
        C. opened first and closed first
        D. none of the above

_D_   15. A lockout/tagout must be removed _____.
        A. by the worker who installed it
        B. by authorized personnel
        C. in accordance with written lockout/tagout procedures
        D. all of the above

Hazardous Material Container Labeling – RTK Labeling

_D_   1. Health hazards

_F_   2. Signal word

_A_   3. Chemical or common name

_E_   4. Physical hazards

_B_   5. Handling and storage instructions

_C_   6. First aid procedures for exposure or contact

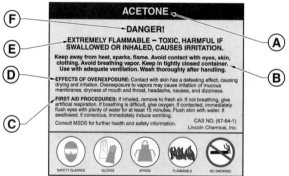

TESTING
TEST 1

12

Name _____ Date _____

True-False

(T) F   1. All steam boilers must have one or more safety valves.

T (F)   2. A hydrostatic test determines whether a steam boiler has sufficient relieving capacity.

(T) F   3. The main steam stop valve, bottom blowdown valves, and feedwater valves should be locked closed before anyone enters the boiler drum.

T (F)   4. Scale protects the boiler heating surfaces.

T (F)   5. Boiler corrosion cannot be prevented as long as water is in contact with metal.

(T) F   6. Water begins to boil at approximately 212°F.

(T) F   7. Fuels most commonly used in boilers are fuel oil, gas, and coal.

T (F)   8. Safety valves on steam boilers are designed to open slowly.

(T) F   9. The steam pressure gauge must be connected to the highest part of the steam side of the boiler.

(T) F   10. With an NOWL in a boiler, the gauge glass is approximately half full.

(T) F   11. A high surface tension on top of the boiler water can lead to foaming.

Identify the ASME Code Symbol Stamps.

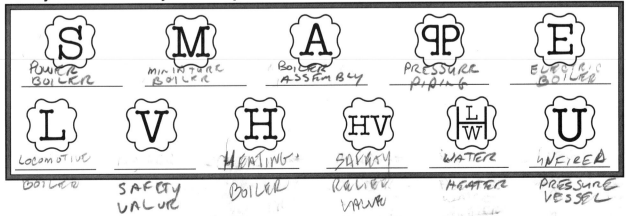

S — POWER BOILER
M — MINIATURE BOILER
A — BOILER ASSEMBLY
PP — PRESSURE PIPING
E — ELECTRIC BOILER
L — LOCOMOTIVE BOILER
V — SAFETY VALVE
H — HEATING BOILER
HV — SAFETY RELIEF VALVE
LW — WATER HEATER
U — UNFIRED PRESSURE VESSEL

89

Multiple Choice

_A_   1. The bottom blowdown on a boiler _____.
     A. removes sludge and sediment from the mud drum
     B. reduces boiler steam pressure
     C. adds makeup water to the boiler
     D. increases boiler priming

_B_   2. A boiler fusible plug is brass or bronze with a core of _____.
     A. zinc
     B. tin
     C. titanium
     D. iron oxide

_B_   3. Spalling in a boiler refers to _____.
     A. hairline cracks in the steam drum
     B. hairline cracks in the refractory
     C. slugs of water in the steam
     D. water in the fuel oil

_B_   4. Fires caused by spontaneous combustion are most likely to occur in the _____.
     A. fuel oil tank
     B. oily waste rag storage
     C. boiler furnace
     D. boiler ash pit

_D_   5. The _____ are a secondary means of determining boiler water level.
     A. water columns
     B. gauge glass blowdown valves
     C. high and low water alarms
     D. try cocks

_C_   6. A siphon installed between the boiler and the pressure gauge protects the Bourdon tube from _____ reaching the pressure gauge.
     A. water
     B. gases of combustion
     C. steam
     D. fuel

_C_   7. Boiler feedwater is chemically treated to _____.
     A. increase circulation
     B. increase oxygen concentration
     C. prevent formation of scale
     D. decrease boiler makeup water

8. In a firetube boiler, soot accumulates on the _____. **[A]**
    A. inside tube surface
    B. outside tube surface
    C. waterwall surface
    D. lowest part of the water side

9. The pressure applied on the boiler during a hydrostatic test should be _____ times the MAWP. **[A]**
    A. 1½
    B. 2
    C. 2½
    D. 3

10. As fuel oil is heated, its viscosity is _____. **[C]**
    A. the same
    B. increased
    C. decreased
    D. ignited

11. The induced draft fan is located between the boiler and the _____. **[D]**
    A. feedwater pump
    B. condensate return tank
    C. evaporator
    D. chimney

12. The flash point of fuel oil is the minimum temperature at which the fuel will _____. **[C]**
    A. support combustion
    B. no longer flow
    C. flash when exposed to an open flame
    D. have its highest Btu content

13. The fire point of fuel oil is the minimum temperature at which the fuel oil will _____. **[A]**
    A. burn continually
    B. no longer flash
    C. flash when exposed to an open flame
    D. have its highest Btu content

14. The pour point of fuel oil is the _____ temperature at which fuel oil will _____. **[B]**
    A. lowest; burn
    B. lowest; flow
    C. highest; burn
    D. highest; flow

15. Fuel oil with a low flash point is _____. **[B]**
    A. used with high pressure boilers
    B. dangerous to handle
    C. only used in low pressure plants
    D. heated to increase its viscosity

A 16. The operating range of a steam boiler is controlled by a(n) _____.
   A. pressure control
   B. air flow interlock
   C. flame scanner
   D. programmer

C 17. Water hammer in steam lines is caused by _____.
   A. low steam pressure
   B. high steam pressure
   C. condensate in the line
   D. a sudden drop in plant load

D 18. A manometer measures _____.
   A. flue gas temperature
   B. volume of air flow
   C. atmospheric pressure
   D. difference in pressure between two points

C 19. The best time to blow down a boiler is when it is _____.
   A. at its peak load
   B. at 75% of its peak load
   C. at its lowest load
   D. being taken out of service

C 20. Boilers that are laid up dry have trays of _____ put in the steam and water drums to absorb moisture.
   A. calcium chloride
   B. hot soda lime
   C. silica gel
   D. potash

A 21. To obtain complete combustion of a fuel, _____, _____, _____, and _____ are required.
   A. mixture; atomization; temperature; time
   B. CO_2; O_2; CO; CO_3
   C. turbulence; feedwater; air; flue gas
   D. hot refractory; air; draft; gases of combustion

B 22. Oxygen in the boiler causes _____.
   A. scale
   B. pitting
   C. foaming
   D. carryover

D _____ 23. When the boiler low water alarm is ringing, the operator should _____.
   A. call the chief engineer
   B. increase the firing rate
   C. decrease the feed water
   D. secure the fires

C _____ 24. A furnace explosion can be prevented by _____.
   A. checking the water level once a shift
   B. testing the safety valves once a month
   C. purging the furnace after ignition failure
   D. testing the safety valve regularly

A _____ 25. A lead sulfide cell, which is used for flame detection, is sensitive to _____.
   A. infrared light
   B. light
   C. heat
   D. temperature increases

D _____ 26. A pop safety valve is a _____ type.
   A. hand and lever
   B. pin and disc
   C. deadweight
   D. spring-loaded

B _____ 27. To test a low water fuel cutoff using an evaporation test, the boiler operator must _____.
   A. close the feedwater stop valve
   B. secure all of the feedwater going to the boiler
   C. completely drain the regulator chamber
   D. have an inspector on hand

B _____ 28. The _____ Hazard Signal System provides container label information about hazardous material.
   A. ASME
   B. NFPA
   C. EPA
   D. ANSI

A _____ 29. To prevent a vacuum from forming on a boiler that is coming off-line, _____.
   A. open the boiler vent
   B. blow down the boiler
   C. dump the boiler
   D. pop the safety valve

C_____ 30. Low pressure boilers equipped with quick-opening and screw valves are blown down by opening the _____ valve first and closing it _____.
   A. quick-opening; first
   B. screw; last
   C. quick-opening; last
   D. quick-opening valves are never used on low pressure boilers

D_____ 31. The viscosity of fuel oil is the measurement of the _____ of the fuel oil.
   A. Btu content
   B. fire point
   C. flash point
   D. internal resistance to flow

D_____ 32. Feedwater is treated chemically before it enters the boiler to _____.
   A. prevent foaming
   B. eliminate blowing down the boiler
   C. change the scale-forming salts to a sludge that will raise the boiler water temperature
   D. change scale-forming salts to a nonadhering sludge

B_____ 33. Complete combustion is defined as burning all fuel using _____.
   A. the theoretical amount of air
   B. a minimal amount of excess air
   C. no excess air
   D. CO_2 and CO

B_____ 34. When testing the safety valve by hand, there should be at least _____ psi of pressure on the boiler.
   A. 1
   B. 5
   C. 15
   D. 160

B_____ 35. The water column on a low pressure steam boiler _____.
   A. reduces fluctuation in water level to prevent carryover
   B. reduces the turbulence of water in the gauge glass
   C. provides a place to install the bottom blowdown valve
   D. a water column is never used on a low pressure boiler

C_____ 36. Safety valves on low pressure boilers can be tested _____.
   A. only by hand
   B. only by pressure
   C. by hand or by pressure
   D. safety valves should never be tested

C   37. Before laying up a boiler, the boiler operator must _____.
- A. notify the boiler inspector
- B. remove the boiler certificate
- C. thoroughly clean the fire and water sides
- D. only clean the water side because soot acts as an insulator

B   38. If water comes out of the top try cock, the boiler operator must _____.
- A. notify the engineer
- B. blow down the boiler
- C. bypass the try cocks and feed by hand
- D. leave the boiler alone, this is normal

A   39. If steam comes out of the bottom try cock, the boiler operator must _____.
- A. secure the fires to the boiler, allow the boiler to cool slowly, and notify the boiler inspector
- B. bypass the try cocks and feed by hand
- C. leave the boiler alone, this is normal
- D. start up a new feedwater pump

D   40. A scotch marine boiler is a _____ boiler.
- A. watertube
- B. cast iron
- C. cast iron firetube
- D. firetube

D   41. When performing a hydrostatic test on a boiler, _____.
- A. all safety valves are removed or gagged
- B. the boiler vent is closed
- C. the main steam stop valve is closed
- D. all of the above

A   42. _____ is the difference in pressure between two points of measurement that causes air or gases to flow.
- A. Draft
- B. Feedwater
- C. Condensate
- D. Carryover

B   43. No. 6 fuel oil has _____ viscosity when compared to No. 2 fuel oil.
- A. no
- B. high
- C. medium
- D. low

B   **44.** _____ is heat transfer that occurs when molecules in a material are heated and the heat is passed from molecule to molecule through the material.
- A. Combustion
- B. Conduction
- C. Condensation
- D. none of the above

D   **45.** Confined space is _____.
- A. large enough for an employee to enter
- B. not designed for continuous employee access
- C. restricted in means for entry and exit
- D. all of the above

TESTING
TEST 2

Name_____ Date_____

True-False

(T) F 1. To prevent water hammer, water should be removed from all steam lines.

T (F) 2. Pressure on water does not change the boiling point of the water.

T (F) 3. A Bourdon tube in a steam pressure gauge is filled with live steam when it is operating.

(T) F 4. A steam trap is a device that removes air and condensate without loss of steam.

(T) F 5. Treated water is used in a boiler to prevent pitting of the boiler metal.

T (F) 6. When dumping a boiler, there should be at least 2 lb of steam in the boiler to force water from the boiler.

(T) F 7. The boiler vent should be open when filling the boiler with water.

(T) F 8. The pressure control must be mounted in a vertical position to ensure accurate operation.

T (F) 9. A burner should always start up in high fire to ensure enough fuel for ignition.

(T) F 10. Carrying too high a water level in the boiler could lead to water hammer.

Multiple Choice

C _____ 1. On boilers that have both quick-opening and screw blowdown valves, the quick-opening valve must be located _____.
  A. after the screw valve
  B. at the NOWL
  C. closest to the boiler and followed by the screw valve
  D. on the boiler vent line

D _____ 2. Entering a boiler for maintenance may require a(n) _____.
  A. OSHA certification
  B. ASME license
  C. boiler operator license
  D. confined space permit

B 3. The Bourdon tube in a steam pressure gauge is protected by a _____.
   A. steam trap
   B. siphon
   C. steam strainer
   D. stopcock

C 4. Draft is measured with a _____.
   A. pyrometer
   B. hydrometer
   C. manometer
   D. pressure gauge

C 5. A steam boiler is blown down to _____.
   A. lower the oxygen level
   B. test the safety valve
   C. remove sludge and sediment
   D. clean the feedwater lines

A 6. The NOWL in a steam boiler is indicated when _____.
   A. steam and water flow out of the middle try cock
   B. water flows out of the top try cock
   C. steam flows out of the bottom try cock
   D. steam and water flow out of the top try cock

B 7. A compound gauge indicates _____.
   A. differential pressure
   B. pressure or vacuum
   C. absolute pressure
   D. the sum of the pressure on two boilers

B 8. Atomization of fuel oil in a rotary cup burner is caused by the _____.
   A. rotating cup
   B. rotating cup and primary air
   C. pressure of the fuel oil
   D. secondary and primary air

A 9. The try cocks on a water column are used _____.
   A. as a secondary means of determining the water level
   B. to blow down the water column
   C. for testing the flame scanner
   D. to remove impurities from the surface of the water

A 10. A low pressure steam boiler has a maximum allowable working pressure (MAWP) of up to _____ psi.
   A. 15
   B. 35
   C. 100
   D. low pressure boilers have no MAWP

A _____ 11. In a firetube boiler, the heat and gases of combustion pass _____.
   A. through the tubes
   B. around the tubes
   C. only through the combustion chamber
   D. both A and B

C _____ 12. Safety valve connections must be approved by the _____.
   A. fire department
   B. shift foreman
   C. ASME
   D. ABMA

D _____ 13. Boiler fittings are necessary for _____.
   A. safety
   B. efficiency
   C. cosmetic purposes
   D. both A and B

D _____ 14. The most important fitting on a boiler is the _____.
   A. low water fuel cutoff
   B. feedwater regulator
   C. superheater
   D. safety valve

C _____ 15. A safety valve _____.
   A. controls the boiler operating range
   B. controls high or low water
   C. prevents the boiler from exceeding its MAWP
   D. controls high and low fire

B _____ 16. Safety valves are designed to _____.
   A. open slowly to prevent water hammer
   B. pop open
   C. open only by hand
   D. open slowly and then close with no drop in pressure

D _____ 17. A steam pressure gauge is calibrated in _____.
   A. pounds per cubic inch
   B. inches of water column
   C. pounds per temperature increase
   D. pounds per square inch

B _____ 18. Bottom blowdown lines on a watertube boiler are located on the _____.
   A. feedwater pump
   B. bottom of the mud drum
   C. fuel oil pump inlet
   D. lowest part of the combustion chamber

A   19. When blowing down a boiler equipped with a quick-opening valve and a screw valve, the quick-opening valve is _____.
- A. opened first and closed last
- B. opened last and closed first
- C. opened first and closed first
- D. opened as operator prefers

A   20. If a quick-opening valve is used as a bottom blowdown valve, it must be located _____.
- A. between the boiler and the screw valve
- B. furthest from the shell of the boiler
- C. cannot be used on low pressure boilers
- D. none of the above

A   21. Surface tension on the water in the steam and water drum is increased by _____.
- A. impurities that float on the surface of the water
- B. inexperienced boiler operators
- C. very soft water
- D. a high steam load

B   22. The surface blowdown line on a boiler is located at the _____.
- A. MAWP
- B. NOWL
- C. lowest heating surface
- D. bottom try cock

A   23. Draft is defined as a difference in pressure that causes _____.
- A. air or gases to flow
- B. a balance of pressure
- C. a back pressure
- D. none of the above

C   24. Natural draft is produced by a(n) _____.
- A. forced draft fan
- B. induced draft fan (used in larger plants)
- C. difference in temperature of a column of gas inside the chimney from a column of air outside the chimney
- D. clean coal bed

C   25. A draft fan located between the boiler and chimney is used in a(n) _____ draft system.
- A. natural
- B. forced
- C. induced
- D. combination forced and induced

__D__ 26. The purpose of the fuel oil return line is to _____.
   A. circulate fuel oil during warm-up
   B. return fuel oil that bypasses the burner
   C. return fuel oil from the relief valve
   D. all of the above

__A__ 27. Duplex strainers are found on _____ of the fuel oil pump.
   A. the suction side
   B. the discharge side
   C. both the suction and discharge sides
   D. none of the above

__D__ 28. Fuel oil heaters must be used when burning No. _____ fuel oil.
   A. 1
   B. 2
   C. 4
   D. 6

__B__ 29. The _____ oversees adherence to codes involving the construction and repairs of boilers and pressure vessels.
   A. EPA
   B. National Board
   C. OSHA
   D. boiler inspector

__D__ 30. The _____ valve controls the firing rate in a high pressure gas system.
   A. solenoid
   B. manual reset
   C. check
   D. butterfly

__A__ 31. The gas pressure regulator in a low pressure gas system reduces the gas pressure to _____ psi.
   A. 0
   B. 5
   C. 10
   D. 15

__B__ 32. Boilers are equipped with a combination gas/fuel oil burner for _____.
   A. greater efficiency
   B. more flexible operation
   C. safer operation
   D. higher rates of combustion

B   33. Draft is measured in _____.
- A. pounds per square inch
- B. inches or tenths of an inch of a vertical water column
- C. inches or tenths of an inch of mercury
- D. ounces per square inch

D   34. _____ draft uses a fan before and after the boiler.
- A. Forced
- B. Induced
- C. Natural
- D. Combination forced and induced

B   35. The amount of draft available in a natural draft system is dependent on the _____.
- A. water pressure
- B. height of the chimney
- C. fly ash in the gases of combustion
- D. types of fans used

C   36. Mechanical draft is produced by the _____.
- A. height of the chimney
- B. diameter of the chimney
- C. power-driven fans
- D. steam jets

D   37. Mechanical draft can be classified as _____.
- A. pressurized
- B. natural
- C. regenerative
- D. forced or induced

A   38. The induced draft fan is located _____.
- A. between the boiler and the chimney
- B. at the base of the chimney
- C. at the burner
- D. in the first pass of the gases of combustion

D   39. Mechanical draft can be used when burning _____.
- A. fuel oil
- B. coal
- C. gas
- D. all of the above

B   40. The type of draft used in a fireplace is _____ draft.
- A. mechanical
- B. natural
- C. forced
- D. combination

B

41. Anthracite coal has a high _____ content.
    A. lignite
    B. fixed carbon
    C. moisture
    D. volatile

C

42. When the temperature of fuel oil is raised, its viscosity _____.
    A. remains the same
    B. is raised
    C. is lowered
    D. cannot be affected by heat

A

43. The temperature at which fuel oil gives off a vapor that ignites readily when exposed to an open flame is its _____ point.
    A. flash
    B. fire
    C. pour
    D. viscosity

C

44. The lowest temperature at which fuel oil will flow is its _____ point.
    A. flash
    B. fire
    C. pour
    D. viscosity

C

45. A(n) _____ is burner control equipment that monitors the burner start-up sequence and the main flame during normal operation.
    A. flue gas analyzer
    B. modulation control system
    C. flame safeguard system
    D. none of the above

TESTING
TEST 3

Name_____ Date_____

True-False

T (F) 1. Blowdown of the boiler is not required if proper feedwater treatment is used.

(T) F 2. Try cocks on a water column can be used to determine the water level if the gauge glass is broken.

(T) F 3. The low water fuel cutoff shuts off the fuel to the burner when a low water condition exists.

(T) F 4. The water column reduces water turbulence to allow a more accurate reading in the gauge glass.

T (F) 5. Blowing down the water column and gauge glass too frequently can cause a false water level.

(T) F 6. A stop valve on the feedwater line closest to the shell of the boiler allows repair of the check valve without dumping the boiler.

(T) F 7. The vacuum pump is designed to discharge air and pump water.

T (F) 8. Makeup water is added when the boiler water is above the NOWL.

(T) F 9. Most makeup water contains some scale-forming salts.

(T) F 10. A globe valve should never be used as a main steam stop valve.

(T) F 11. When open, gate valves offer no restriction to flow.

Multiple Choice

__D__ 1. The purpose of the ON/OFF pressure control is to _____.
  A. open and close the water column
  B. regulate air flow
  C. regulate fuel flow
  D. start and stop the burner on steam pressure demand

__A__ 2. A steam siphon ensures that _____ does not enter the boiler pressure control.
  A. steam
  B. water
  C. air
  D. gas

105

__D__ 3. The flame scanner is sensitive to _____.
   A. heat
   B. temperature
   C. pressure
   D. infrared rays

__B__ 4. In the event of a flame failure, the programmer _____.
   A. sends more fuel to start a new firing cycle
   B. secures the fuel and purges the furnace
   C. vents the combustion chamber
   D. starts the induced draft fan

__D__ 5. Pressure gauges are calibrated in pounds per _____.
   A. square foot
   B. cubic inch
   C. vertical inch
   D. square inch

__C__ 6. The boiler steam pressure gauge must be connected to the _____ of the boiler.
   A. lowest part of the steam side
   B. lowest part of the water side
   C. highest part of the steam side
   D. highest part of the water side

__A__ 7. Vacuum gauges are calibrated in inches of _____ atmospheric pressure.
   A. mercury below
   B. water below
   C. mercury above
   D. water above

__A__ 8. Boiler water must be treated to prevent _____.
   A. formation of scale
   B. overpressure
   C. flame failure
   D. none of the above

__B__ 9. Oxygen in the boiler causes _____.
   A. scale
   B. pitting of boiler metal
   C. caustic embrittlement
   D. carryover

__B__ 10. Carryover can lead to _____.
   A. excess fuel oil temperature
   B. water hammer
   C. gas overpressure
   D. air in the feedwater lines

11. Oxygen present in the boiler water is commonly removed by _____.
    A. adding lead sulfide
    B. heating the feedwater
    C. using the bottom blowdown valves
    D. using the surface blowdown valves

12. When taking over a shift, the boiler operator first checks the _____.
    A. boiler room log
    B. fuel oil supply
    C. water level on all boilers that are on the line
    D. bottom blowdown valves

13. The purpose of a flame safeguard system is to protect the boiler from _____.
    A. improper fuel oil pressure
    B. a possible furnace explosion
    C. exceeding its MAWP
    D. starting in low fire

14. After verifying the proper boiler water level, the burner may be started after _____.
    A. filling the vacuum tank
    B. venting the drum condenser
    C. purging the furnace
    D. purging the superheater

15. As the boiler is cooling down, the boiler operator must maintain the _____.
    A. NOWL
    B. normal draft condition
    C. normal feedwater temperature
    D. 10% CO_2 reading

16. Boilers that are out of service for an extended period of time _____.
    A. are stored with an NOWL
    B. are filled with oxygen
    C. require more than 8 psi in the boiler
    D. require proper lay-up procedures

17. If there is a danger of the boiler freezing, the boiler should be laid up _____.
    A. with a light fire
    B. by using steam from the header to keep the boiler warm
    C. dry with all water removed
    D. using an antifreeze

18. A low water fuel cutoff _____.
    A. shuts the burner down
    B. increases steam pressure to the load
    C. adds water to the boiler
    D. purges the boiler furnace

A   **19.** A boiler that has had a low water condition should be _____.
- A. thoroughly examined for signs of overheating
- B. thoroughly examined for scale buildup
- C. brought up to full steam pressure to test for leaks
- D. brought up to the MAWP as soon as possible

A   **20.** Flame scanners are _____.
- A. equipped with a flame sensor
- B. found on fuel oil strainers
- C. found only on high pressure boilers
- D. found only on low pressure boilers

D   **21.** A furnace explosion can be caused by _____.
- A. excess draft
- B. a low steam pressure condition
- C. an overheated furnace
- D. an accumulation of fuel vapors

B   **22.** Class _____ fires burn oil, grease, paint, and other flammable liquids.
- A. A
- B. B
- C. C
- D. D

B   **23.** An unsafe condition should be reported to the _____.
- A. shift operator
- B. immediate supervisor
- C. fire department
- D. plant personnel department

D   **24.** To start and sustain a fire, _____ are required.
- A. fuel, heat, and CO_2
- B. fuel, combustible matter, and CO_2
- C. fuel, heat, and nitrogen
- D. fuel, heat, and oxygen

A   **25.** A fire caused by the ignition of wood, paper, or textiles is a Class _____ fire.
- A. A
- B. B
- C. C
- D. D

A   **26.** A _____ is the use of locks, chains, or other physical restraints to prevent the operation of equipment.
- A. lockout
- B. tagout
- C. permit authorization
- D. restriction form

27. The low water fuel cutoff should be blown down _____. [A]
    A. daily or more often
    B. weekly
    C. monthly
    D. during annual inspection

28. The _____ is sensitive to water temperature. [B]
    A. low water fuel cutoff
    B. aquastat
    C. vaporstat
    D. feedwater regulator

29. A drop in water level in a steam boiler causes the automatic feedwater regulator to _____. [B]
    A. close the fuel oil solenoid valve
    B. increase the flow of feedwater
    C. reduce the flow of feedwater
    D. increase the fuel oil to the burner

30. A _____ is a regulation or minimum requirement. [B]
    A. standard
    B. code
    C. technical bulletin
    D. recommendation

31. In the safe operation of a steam boiler, the most important rule to follow is to _____. [D]
    A. fill out the boiler room log
    B. perform bottom blowdowns regularly
    C. read the operation manual daily
    D. maintain the proper boiler water level at all times

32. The purpose of try cocks is to _____. [B]
    A. remove sludge and sediment
    B. determine the water level in the boiler
    C. draw water samples
    D. blow down the gauge

33. A(n) _____ valve is allowed between the boiler shell and the safety valve. [D]
    A. os&y
    B. lever
    C. check
    D. none of the above

34. A low pressure boiler safety valve setting cannot exceed _____ psi. [A]
    A. 15
    B. 20
    C. 25
    D. 160

D 35. _____ allow for the movement caused by expansion and contraction of steam lines from heating and cooling.
A. Lagging joints
B. Convection valves
C. Bellow plates
D. Expansion bends

A 36. In a flame safeguard system, before the main flame fuel valve(s) can open, the _____ must first be proven.
A. pilot flame
B. main burner
C. water level
D. feedwater flow

A 37. The low water fuel cutoff should be tested _____.
A. with the burner firing
B. with the burner off
C. by removing the flame scanner
D. once a heating season

A 38. If the low water alarm starts ringing, _____.
A. secure the burner
B. blow down the water column
C. add feedwater quickly and reduce the firing rate
D. start another feedwater pump

A 39. The most accurate method used to determine the amount of feedwater treatment required is _____.
A. boiler water analysis
B. measurement of feedwater pressure
C. measurement of blowdown pressure
D. checking boiler water temperature

A 40. The boiler vent should be kept open when _____.
A. the boiler is warmed up
B. the boiler has maximum line pressure
C. steam is blowing out of the vent
D. the boiler is cut off the line

C 41. To perform an evaporation test on the low water fuel cutoff, _____.
A. have the chief engineer in attendance
B. close the main steam stop valve
C. secure all feedwater and makeup water going to the boiler
D. completely drain the low water fuel cutoff float chamber

TESTING
TEST 4

Name_____ Date_____

True-False

- (T) F 1. Steam boilers that are out of service for an extended period require dry lay-up.
- T (F) 2. The dry method of boiler lay-up requires the boiler to be left open with air circulated to keep the boiler drums and tubes dry.
- T (F) 3. Chemical treatment of boiler water is not required for wet lay-up if the water has been deaerated.
- (T) F 4. The flash point of fuel oil is lower than the fire point.
- (T) F 5. The fire point of fuel oil is the minimum temperature at which fuel oil burns continuously.
- T (F) 6. No. 6 fuel oil burns with a clean flame when it is not heated.
- (T) F 7. If the gasket leaks on a duplex strainer, air could be drawn into the fuel oil lines.
- T (F) 8. Dirty strainers produce low suction readings.
- T (F) 9. Rotary cup burners can only burn No. 2 fuel oil.
- (T) F 10. Air used to atomize fuel oil is primary air.
- T (F) 11. Combination gas/fuel oil burners are used only in high pressure plants.

Multiple Choice

__B__ 1. A _____ should be used to clean the inside of a gauge glass.
  A. wire brush
  B. cloth wrapped around a wooden dowel
  C. screwdriver wrapped with a paper towel
  D. sandpaper mounted on a steel rod

__A__ 2. A _____ is an accepted reference or practice.
  A. standard
  B. code
  C. recommendation
  D. none of the above

111

C_____ 3. An aquastat is an automatic device that is controlled by sensing _____.
   A. water pressure
   B. steam pressure
   C. water temperature
   D. steam temperature

A_____ 4. Sudden extreme changes of temperature in a furnace can cause _____.
   A. spalling
   B. a drop in chimney temperature
   C. increased soot deposits
   D. a rise in chimney temperature

A_____ 5. The number of blowdowns a boiler requires is determined by _____.
   A. boiler water analysis
   B. checking chimney temperature
   C. boiler manufacturer data
   D. steam flow in the plant

B_____ 6. Complete combustion is the combustion of fuel with _____.
   A. the theoretical amount of air
   B. the minimum amount of excess air
   C. no excess air
   D. no smoke or CO_2

C_____ 7. Oxygen in the boiler causes _____.
   A. priming
   B. foaming
   C. pitting
   D. carryover

A_____ 8. No. _____ fuel oil has the highest heating value.
   A. 2
   B. 4
   C. 5
   D. 6

A_____ 9. The amount of _____ draft produced is affected by the temperature outside.
   A. natural
   B. forced
   C. induced
   D. balanced

C_____ 10. To prepare a boiler for inspection, the boiler should be _____.
   A. at MAWP
   B. on-line
   C. cool, open, and thoroughly clean
   D. ready to dump when the inspector arrives

11. Information on the boiler nameplate includes _____. *(D)*
    A. MAWP
    B. date of manufacture
    C. manufacturer
    D. all of the above

12. The safety valve is commonly tested during normal plant operation by the _____. *(C)*
    A. chief engineer
    B. state inspector
    C. operator on duty
    D. plant manager

13. In a straight-tube watertube boiler, heat and gases of combustion pass through the _____. *(B)*
    A. tubes
    B. furnace
    C. fuel oil strainer
    D. all of the above

14. A feedwater pump can become steambound if _____. *(D)*
    A. steam pressure is too high
    B. water pressure is too low
    C. steam pressure is too low
    D. feedwater temperature is too high

15. When burning No. 6 fuel oil, a _____ is needed. *(A)*
    A. fuel oil heater
    B. feedwater heater
    C. air preheater
    D. none of the above

16. The height of the chimney determines the amount of _____ draft. *(D)*
    A. forced
    B. induced
    C. combination forced and induced
    D. natural

17. The water column on a low pressure boiler is used to _____. *(C)*
    A. measure the water level
    B. indicate steam pressure
    C. reduce fluctuation of the water level in the gauge glass
    D. water columns are never used on low pressure boilers

18. To protect against a furnace explosion, _____. *(B)*
    A. keep the water side clean
    B. purge the boiler after all flame failures
    C. check the damper
    D. test safety valves once a week

C 19. A water column should be blown down _____.
   A. once a week
   B. once a month
   C. once a shift
   D. when taking a boiler off-line

D 20. High and low fire of the burner are controlled by a(n) _____.
   A. vaporstat
   B. pressure control
   C. aquastat
   D. modulating pressure control

B 21. A clogged duplex fuel oil strainer located on the suction side of the pump results in _____.
   A. a high discharge pressure
   B. a high reading on the suction gauge
   C. a high fuel oil temperature
   D. increased fuel oil consumption

D 22. A fan located between the breeching and chimney is a(n) _____ fan.
   A. chimney
   B. breeching
   C. forced draft
   D. induced draft

D 23. _____ is when water is carried over from the boiler into the steam lines.
   A. Purging
   B. Chattering
   C. Throttling
   D. Priming

C 24. Impurities on the surface of the water in a steam and water drum are removed by a(n) _____.
   A. continuous blowdown
   B. intermittent blowdown
   C. surface blowdown
   D. bottom blowdown

D 25. A main steam stop valve is most commonly a(n) _____ valve.
   A. check
   B. globe
   C. modulating
   D. os&y gate

_C_   26. The condensate return temperature can be affected by a malfunctioning _____.
- A. feedwater pump
- B. fuel oil pump
- C. steam trap
- D. fuel oil strainer

_C_   27. A pressure control on a steam boiler is used to _____.
- A. control operating fuel oil temperature
- B. control makeup water pressure
- C. control high and low fire
- D. start and stop the breeching cleanout

_B_   28. A feedwater pump may become steambound if _____.
- A. steam pressure overcomes water pressure
- B. the temperature of the feedwater in the heater is too high
- C. air enters the suction side of the pump
- D. pressure in the pump is too low

_A_   29. Before dumping a boiler, _____.
- A. allow the boiler to cool
- B. call the inspector
- C. open the blowdown valve with 5 psi to 10 psi on the boiler
- D. open all surface blowdown valves

_C_   30. The amount of blowdowns a boiler requires is determined by _____.
- A. the state inspector
- B. the boiler operator in charge
- C. boiler water analysis
- D. the boiler manufacturer

_D_   31. Foaming of water results when the water is contaminated with foreign material that causes _____.
- A. an increase in makeup water added
- B. more blowdowns
- C. an increase in fuel consumption
- D. an increase in surface tension

_B_   32. Blowing down is most effective when the steam output is _____.
- A. at a high rate
- B. at a low rate
- C. at zero
- D. early in the shift

_D_   33. The water column must be located _____.
- A. on the right side of the boiler
- B. on the left side of the boiler
- C. on the front of the boiler
- D. at the NOWL

C   34. The low water fuel cutoff should be tested with an evaporation test _____.
- A. daily
- B. weekly
- C. monthly
- D. yearly

C   35. A pressure control is protected from live steam by a _____.
- A. stop valve
- B. try cock
- C. siphon
- D. check valve

B   36. A boiler produces black smoke when there is _____.
- A. low atmospheric pressure
- B. an improper mixture of air and fuel
- C. excess secondary air
- D. excess primary air

C   37. In order to properly test the low water fuel cutoff, _____.
- A. the burner must be OFF
- B. there must be no pressure on the boiler
- C. the burner must be firing
- D. the fuel must be shut OFF

A   38. The purpose of an expansion tank in a hot water heating system is to allow for the expansion of _____.
- A. water
- B. hot air
- C. air and steam
- D. gas in the burner

C   39. The flame scanner is located on the _____.
- A. main steam line
- B. fuel oil line
- C. front of the furnace
- D. bottom blowdown line

B   40. The pressure control controls the boiler operating range by _____.
- A. regulating the fuel oil pressure
- B. starting and stopping the burner
- C. changes in water temperature
- D. causing the safety valve to relieve pressure

A   41. A heavy accumulation of soot on a boiler heating surface results in _____.
- A. loss of boiler efficiency
- B. increased heat transfer
- C. loss of fire
- D. safety valve popping

TESTING
TEST 5

Name_____ Date _____

True-False

T F 1. A low water condition is a common cause of boiler room accidents.
T **F** 2. A steam pressure gauge is calibrated in pounds per square foot.
T F 3. The Bourdon tube of a steam pressure gauge must be protected from live steam.
T **F** 4. A bottom blowdown line returns water to the boiler.
T F 5. The functions of a steam trap are to remove air and condensate without the loss of steam.
T F 6. Purging a furnace before firing can prevent a furnace explosion.
T F 7. Boilers burning soft coal require a large furnace volume to complete combustion.
T F 8. Smoke is a sign of incomplete combustion.
T F 9. The furnace must be purged after any flame failure.
T **F** 10. An induced draft fan is located on the front of the boiler.
T F 11. When using natural draft, there is a limit to the amount of fuel that can be burned.
T **F** 12. A forced draft fan is located in the breeching.
T **F** 13. Oxygen in the boiler water is used to remove nonadhering sludge.

Multiple Choice

D ___ 1. An ignition failure results in the operation of the _____.
   A. low water fuel cutoff
   B. feedwater makeup valve
   C. high fire control
   D. flame scanner

A ___ 2. An automatic feedwater regulator is used to _____.
   A. ensure the proper water level in the boiler
   B. shut off the burner in the event of low water
   C. control the burner operating range
   D. modulate gas pressure to the burner

B 3. On a steam boiler, testing the operation of the safety valve by hand with the boiler under pressure should be performed _____.
   A. at the start of each heating season
   B. every 30 days
   C. annually
   D. when maximum water pressure is achieved

C 4. In a natural circulation hot water heating system, excess water is collected in the _____.
   A. aquastat
   B. blowdown tank
   C. expansion tank
   D. relief valve tank

B 5. If ignition fails during burner startup, the _____ protects the boiler.
   A. low water fuel cutoff
   B. flame scanner
   C. vaporstat
   D. aquastat

C 6. The _____ is the automatic control that protects a boiler from being fired with a low water condition.
   A. aquastat
   B. whistle valve
   C. low water fuel cutoff
   D. flame scanner

B 7. An indication of incomplete combustion is _____.
   A. increased combustion efficiency
   B. black smoke from the chimney
   C. an increase in water level
   D. higher steam pressure generated

C 8. A(n) _____ pressure control controls the amount of steam produced by changing the firing rate.
   A. safety
   B. ON/OFF
   C. modulating
   D. aquastat

C 9. The _____ operates by sensing boiler water temperature.
   A. flame scanner
   B. vaporstat
   C. aquastat
   D. pressure control

A   10. The purpose of performing a try lever test on a safety valve is to _____.
   A. test valve operation
   B. set the operating range of the boiler
   C. test popping pressure
   D. determine the NOWL

D   11. A feedwater regulator depends on the operation of a _____.
   A. water pressure solenoid
   B. steam pressure orifice
   C. water temperature gauge
   D. float-controlled valve

C   12. Furnace explosions can be prevented by _____.
   A. blowing down the low water fuel cutoff daily
   B. testing the safety valve daily
   C. purging the furnace after any flame failure
   D. none of the above

C   13. To increase _____, the height of the chimney must be increased.
   A. forced draft
   B. induced draft
   C. natural draft
   D. combustion gas

A   14. In most plants, the water column and gauge glass should be blown down at least _____.
   A. each 8-hour shift
   B. weekly
   C. monthly
   D. annually

B   15. When blowing down the low water fuel cutoff, _____.
   A. more fuel is burned
   B. the burner shuts off
   C. steam pressure increases
   D. condensate temperature increases

A   16. The purpose of the aquastat is to _____.
   A. control the operating temperature range of a hot water heating boiler
   B. feed water to the system
   C. maintain a minimum water level in the boiler
   D. provide for the expansion of water

C   17. A pressure-reducing valve in a hot water heating system _____.
   A. maintains a specified water pressure in the boiler
   B. shuts the burner OFF when there is low water in the boiler
   C. reduces makeup water pressure
   D. regulates the flow of steam from the boiler

__B__ 18. A safety relief valve is rated in _____ per hour.
- A. NOWL
- B. Btu
- C. psi
- D. gallons

__B__ 19. Makeup water fed to the boiler is _____.
- A. treated feedwater fed to the burner
- B. water added to the boiler
- C. condensate return water
- D. none of the above

__C__ 20. The water column outlets of a steam boiler are connected to the _____.
- A. mud drum
- B. blowdown line
- C. water and steam section of a steam boiler
- D. water tubes of the boiler

__C__ 21. A low pressure boiler commonly receives makeup water from the _____.
- A. boiler feed pump
- B. injector
- C. automatic city water makeup feeder
- D. return lines

__C__ 22. A pressure control is protected from live steam by a(n) _____.
- A. mercury switch
- B. shutoff valve
- C. siphon
- D. inspector's test valve

__C__ 23. The safety valve is located _____.
- A. in the blowdown line
- B. on the main steam header
- C. on the highest part of the steam side of the boiler
- D. on the try cock

__C__ 24. In most states, the only type of safety valve allowed on steam boilers is the _____ type.
- A. lever
- B. direct-loaded pop
- C. spring-loaded pop-off
- D. vacuum relief

__C__ 25. A stop valve and a check valve are usually placed _____.
- A. in the steam line leaving a watertube boiler
- B. in the blowdown line of a firetube boiler
- C. as close to the boiler as possible on the feedwater line
- D. in line with the pressure control

26. The steam gauge is calibrated in _____. [C]
    A. pounds of pressure
    B. inches of pressure
    C. pounds per square inch
    D. pounds per square foot

27. No. 6 fuel oil _____. [C]
    A. is atomized as it leaves the fuel oil tank
    B. is mixed with No. 2 fuel oil before it is atomized
    C. must be heated before it can be burned
    D. has a lower heating value than No. 2 fuel oil

28. A compound pressure gauge indicates _____. [C]
    A. the difference in pressure
    B. the sum of two line pressures
    C. either pressure or vacuum
    D. three different pressures at the same time

29. At NOWL using try cocks, _____. [D]
    A. water only is discharged out of the top try cock
    B. steam only is discharged out of the bottom try cock
    C. steam and water are discharged out of the bottom try cock
    D. steam and water are discharged out of the middle try cock

30. The purpose of the feedwater regulator is to _____. [B]
    A. sound an alarm if the water level in the boiler is too high
    B. maintain the proper water level in the boiler
    C. shut off the burner if the water gets too low
    D. keep the steam pressure within the safe limits

31. The low water fuel cutoff should be blown down _____. [A]
    A. every day
    B. once a month
    C. at the end of the heating season
    D. once a week

32. When a low water condition occurs, the low water fuel cutoff _____. [C]
    A. increases the feedwater supply to the boiler
    B. sounds an alarm to warn of low water
    C. shuts off the fuel supply to the burner
    D. places the boiler in low fire

33. To obtain an accurate reading of the water level in the gauge glass, _____. [C]
    A. add city water makeup
    B. use the bottom blowdown valve
    C. open the gauge glass blowdown valve wide and then close it
    D. open all try cocks

C_____ 34. To change the operating pressure of a steam boiler, _____.
  A. adjust the fuel supply
  B. open the bottom blowdown valve wide and then close it
  C. adjust the pressure control
  D. look at the gauge glass

B_____ 35. A lead sulfide cell is used in a(n) _____.
  A. pressure control
  B. flame scanner
  C. low water fuel cutoff
  D. evaporation test

D_____ 36. When performing a hydrostatic test on a boiler, the safety valves must be _____.
  A. kept free to pop at a set pressure
  B. opened
  C. plugged at the discharge end
  D. gagged or removed and the opening must be blank flanged

A_____ 37. A feedwater pump is used to _____.
  A. pump water into a boiler
  B. pump out the boiler room sump pit
  C. circulate water through the hot water heating system
  D. pump fuel oil to the fuel oil burner

C_____ 38. The blowdown line commonly discharges to a(n) _____.
  A. sewer
  B. condensate return tank
  C. open sump or blowdown tank
  D. none of the above

D_____ 39. The heating surface of a boiler is _____.
  A. in the furnace
  B. only found on water tube boilers
  C. that part of the boiler where water is located
  D. the part of the boiler where water is on one side and gases of combustion are on the other side

A_____ 40. The range of the steam pressure gauge is _____.
  A. 1½ to 2 times the MAWP
  B. the MAWP
  C. not more than 6% over the MAWP
  D. gauge plus atmospheric pressure

TESTING
TEST 6

Name_____ Date _____

True-False

(T) F   1. In a low pressure gas burner, the gas regulator reduces gas pressure to 0 psi.
(T) F   2. A refrigerant absorbs heat in the evaporator of a compression refrigeration system.
T   (F) 3. The range of a steam pressure gauge should be 2½ to 3 times the MAWP of the boiler.
T   (F) 4. The temperature and pressure of a refrigerant in a compression refrigeration system is lowered when it is compressed.
(T) F   5. An absorption refrigeration system commonly uses ammonia as the refrigerant.
(T) F   6. Sensible heat is the amount of heat that changes the measurable temperature of a substance but not its state.
(T) F   7. Mechanical energy can be changed to heat energy in a refrigeration system.
T   (F) 8. An aquastat is used on a compression tank to control flow to the circulating pump.
(T) F   9. A Scotch marine boiler is a type of firetube boiler.
T   (F) 10. A pressure control regulates the firing of the burner based on condensate return flow.
T   (F) 11. A feedwater check valve is commonly installed between the stop valve and the boiler.
(T) F   12. In colder climates, No. 6 fuel oil requires the use of tank heaters.
T   (F) 13. Cooling systems are commonly rated in Btu of cooling per hour.
T   (F) 14. Perfect combustion occurs when a boiler has no soot in the gases of combustion.
(T) F   15. Scale is caused by an accumulation of minerals present in boiler water.

Multiple Choice

__B__   1. Combustion efficiency in the burner is controlled by _____.
         A. primary air
         B. secondary air
         C. forced combustion gas
         D. all of the above

__C__ 2. In a compression system, liquid refrigerant under high pressure is allowed to drop in pressure by the _____.
   A. absorber
   B. absorbent
   C. expansion valve
   D. condenser

__C__ 3. The capacity of a safety valve is measured by the amount of steam that can be discharged per _____.
   A. shift
   B. minute
   C. hour
   D. blowdown

__C__ 4. Heat added to a substance that changes its state without a change in temperature is _____ heat.
   A. super
   B. sensible
   C. latent
   D. mechanical

__C__ 5. In a refrigeration system, heat is _____ when a fluid changes from a gas to a liquid.
   A. decreased
   B. absorbed
   C. released
   D. compressed

__B__ 6. A(n) _____ is printed material used to relay chemical hazard information from the manufacturer to the employee.
   A. OSHA regulation
   B. MSDS
   C. EPA specification sheet
   D. ASME notice

__B__ 7. Excessive water in the boiler can lead to _____.
   A. scale deposits
   B. water hammer
   C. flame failure
   D. all of the above

__D__ 8. The _____ is a government regulatory agency that was established to control and abate pollution.
   A. DOT
   B. OSHA
   C. ANSI
   D. EPA

B   9. An absorption cooling system does not include a(n) _____.
- A. condenser
- B. compressor
- C. evaporator
- D. generator

C   10. In a lithium bromide and water cooling system, refrigerant is heated with a steam coil in the _____.
- A. evaporator
- B. compressor
- C. generator
- D. condenser

D   11. A Bourdon tube is used in a(n) _____.
- A. compression tank
- B. evaporator
- C. compressor
- D. none of the above

D   12. Air used in the combustion process is classified as _____ air.
- A. primary
- B. secondary
- C. excess
- D. all of the above

A   13. In the low pressure zone of a compression refrigeration system, _____ by the refrigerant.
- A. heat is absorbed
- B. steam pressure is increased
- C. lithium bromide is produced
- D. ammonia vapors are generated

B   14. Soot buildup on heating surfaces acts as an _____ to prevent the transfer of heat.
- A. evaporator
- B. insulator
- C. ionizer
- D. all of the above

D   15. Absorption cooling systems commonly use combinations of _____ and water as a refrigerant.
- A. ammonia
- B. lithium chloride
- C. lithium bromide
- D. all of the above

A 16. Flame scanners can use a _____ to sense infrared rays of the pilot light and main burner.
   A. lead sulfide cell
   B. solenoid
   C. purge sensor
   D. all of the above

D 17. In a compression refrigeration system, high pressure vapor is converted from a gas to a liquid in the _____.
   A. evaporator
   B. compressor
   C. generator
   D. condenser

C 18. Oxygen is removed from the boiler water by _____ the water.
   A. adding minerals to
   B. chilling
   C. heating
   D. all of the above

A 19. In a compression refrigeration system, the refrigerant absorbs heat in the _____.
   A. diverter valve
   B. compressor
   C. generator
   D. none of the above

C 20. Steam that has released its heat turns to _____.
   A. evaporated steam
   B. sensible steam
   C. condensate
   D. latent vapors

A 21. A low water fuel cutoff should be tested _____.
   A. daily
   B. weekly
   C. monthly
   D. annually

D 22. Some _____ containing chlorofluorocarbons (CFCs) are believed to cause damage to the earth's ozone layer.
   A. lithium bromides
   B. nitrogen fuels
   C. steam generators
   D. refrigerants

Testing 127

__C__ 23. The burning of all of the fuel in the burner using the minimum amount of air is _____ combustion.
   A. perfect
   B. theoretical
   C. complete
   D. incomplete

__A__ 24. The temperature at which fuel oil must be heated to burn continuously when exposed to an open flame is the _____ point.
   A. fire
   B. flash
   C. ignition
   D. thermal

__C__ 25. Water used as a medium in indirect cooling systems must be kept above _____ °F.
   A. 0
   B. 22
   C. 32
   D. 212

__B__ 26. Combustibles in a Class A fire are _____.
   A. grease and gasoline
   B. paper and wood
   C. paints and solvents
   D. electrical equipment

__C__ 27. The furnace must be _____ after every flame failure.
   A. cleaned
   B. blown down
   C. purged
   D. pressurized

__D__ 28. The heating value of a fuel oil is expressed in _____.
   A. tons of heating
   B. therms
   C. flame units
   D. British thermal units

__A__ 29. Hard or _____ coal produces less smoke than soft coal.
   A. anthracite
   B. bituminous
   C. volatile
   D. none of the above

30. A low pressure steam boiler has an MAWP of _____ psi.
    A. 5
    B. 10
    C. 25
    D. none of the above

Answer: B

31. There are no tubes in a _____ boiler.
    A. cast iron sectional
    B. vertical wet-top
    C. Scotch marine
    D. multiple-pass dry-top

Answer: A

32. Pressure in the high pressure side of a compression refrigeration system is produced by the _____.
    A. expansion valve
    B. condenser
    C. compressor
    D. evaporator

Answer: C

33. In a refrigeration system, heat is _____ when a fluid changes from a liquid to a gas.
    A. produced
    B. absorbed
    C. released
    D. all of the above

Answer: B

34. The most common medium used to transport heat from an area to be cooled is _____.
    A. water
    B. ammonia
    C. Freon and water
    D. all of the above

Answer: A

35. Boiler fittings are manufactured in accordance with the _____ code.
    A. boiler certification
    B. ASME
    C. NFPA
    D. OSHA

Answer: B

36. Boiler water level can be determined using the _____ if the gauge glass is broken.
    A. pressure gauge
    B. siphon gauge
    C. try cocks
    D. leveling stick

Answer: C

37. A _____ blowdown is performed to reduce foaming of the boiler water.
    A. bottom
    B. surface
    C. priming
    D. huddling

38. Hot water boilers operating with a 250°F water temperature and _____ psi water pressure or less are classified as low pressure.
    A. 15
    B. 100
    C. 160
    D. 212

39. Oxygen in boiler water causes _____.
    A. corrosion
    B. rusting
    C. pitting
    D. all of the above

40. A(n) _____ control system controls the amount of steam produced by changing the burner firing rate.
    A. modulating
    B. ON/OFF
    C. proving
    D. indirect

41. Safety valves are repaired by _____.
    A. the operator as required
    B. an authorized manufacturer representative
    C. the boiler inspector
    D. the plant manager

TESTING
TEST 7

Name_____ Date _____

Essay

1. What is the first thing a boiler operator must do when taking over a shift?
2. What steps must be taken if steam comes out of the bottom try cock?
3. What steps must be taken if water comes out of the top try cock?
4. What are the different ways of getting water into the boiler?
5. What is the function of a pressure control?
6. How are safety valves tested?
7. How often should safety valves be tested?
8. What is the function of a low water fuel cutoff?
9. How is a low water fuel cutoff tested?
10. How often should a safety valve pop on a low pressure steam boiler?
11. At what pressure should a safety valve pop on a low pressure steam boiler?
12. What is meant by purging the furnace?
13. What is the most important valve on the boiler?
14. How often should a boiler have a bottom blowdown?
15. How is the flame scanner tested?
16. What methods can be used to determine the water level in the boiler?
17. What is the main cause of smoke?
18. What are the different types of draft used in boilers?
19. What is the function of a check valve on the feedwater line?
20. What is the function of a stop valve on the feedwater line?
21. What actions would be required if the safety valve was popping and the steam pressure indicated 30 psi on the boiler?
22. What is meant by perfect, complete, and incomplete combustion?
23. What could cause a furnace explosion?
24. What are the steps necessary for preparing for a boiler inspection?
25. How often should a gauge glass be blown down?
26. What is the difference between a forced circulation hot water heating system and a natural circulation hot water heating system?
27. What procedure is followed when performing a hydrostatic test on a boiler?

28. What does the abbreviation NOWL stand for?
29. What does the abbreviation MAWP stand for?
30. What is the cause of carryover?
31. What is the difference between natural and mechanical draft?
32. What causes a feedwater pump to become steambound?
33. What is the difference between a fast gauge and a slow gauge?
34. What is the cause of foaming in the boiler?
35. What is the difference between a gate valve and a globe valve?
36. What is the difference between a watertube and a firetube boiler?
37. Where should a water column be located?
38. When is the best time to give a boiler a bottom blowdown?
39. What is the function of a boiler vent?
40. What is used to control high fire or low fire in a boiler?
41. How can a boiler operator tell when an os&y valve is in the open position?
42. What is the function of the vacuum pump?
43. How often should the automatic city water makeup feeder be blown down?
44. What are the most commonly used types of steam traps?
45. What are two methods of testing for the proper function of a steam trap?
46. How often should the fuel oil strainer be cleaned?
47. What are two commonly used types of fuel oil burners?
48. What are the advantages of using a combination gas/fuel oil burner?
49. What is meant by primary, secondary, and excess air?
50. What is the function of a flame safeguard system?
51. What is hydrostatic pressure?
52. What are three types of heat transfer that occur in a boiler?
53. What is the function of a backflow preventer?
54. Why are expansion bends sometimes required on steam lines?
55. What is an evaporation test?
56. What are the benefits of a boiler room log?
57. What is the function of an aquastat?
58. What is a compression refrigeration system?
59. What special safety precautions must be followed when entering a confined space?
60. How is a boiler operator license obtained?